Tree generation and enumeration

Jesse Sakari Hyttinen

Tree generation and enumeration

A collection of mathematical ideas in graph theory

Publisher: BoD – Books on Demand, Helsinki, Finland

Producer: BoD – Books on Demand, Norderstedt, Germany

ISBN: 978-952-80-6122-9

For my father

Contents

0.Introduction

Mathematics. A beautiful, elegant branch of study and research – and the source of my creativity. No matter how more sophisticated and precise the theories in physics evolve to, a proven theorem in mathematics stays true no matter what. That is the true strength of this discipline. A wonderful thing about mathematics is also the fact that it is a collaboration of generations of mathematicians spanning a time of over two millennia – easily more than tens of thousands of papers in pure and applied mathematics.

As an international, scientific language, this discipline has the capability to express aspects of physics, chemistry, engineering sciences, economics and even biology. The modern technology we have owes a lot to scientists, and these scientists have used – you guessed it – mathematics, to derive their sophisticated theories and models, benefitting the societies as we know it. This book is my contribution to mathematics. Whether it has any application areas outside mathematics, such as computer science, it still contains a theory for trees, and in a way connects trees to partitions and rooted trees to

Peano axioms, thus showing a link between number theory, mathematical logic, and graph theory.

I hope this book inspires you, and mathematicians, students, teachers, enthusiasts, and hobbyists as well. As this book was made for you all and may hopefully give you an insight into how mathematics is full of opportunities, a pinnacle of creativity in human endeavors.

If you would like to know how mathematics could be a form of art and a competitive game as well, then check out the Treespeak section. The goal is to create a community for theorists, also known as treespeak users, whether it be a subreddit or the content feed of my Instagram account. Theorists can hone their skills in partition form treespeak with the game I have created, and a future goal would be to expand this into symbolic treespeak, and a global, competitive game. Another goal could be to create an organization that manages and examines the competition setting of treespeak. This organization would also research how the knowledge of treespeak would evolve through time and could establish prizes for accomplished people in treespeak and related subjects found in this book.

1. The secret language of trees

In the year 2014 I started to research one particular problem in graph theory. The problem was from the film Good Will Hunting:

Draw all homeomorphically irreducible trees of size n = 10.

So, one had to draw all the size 10 series-reduced free trees. May sound complicated, but it actually is quite simple. The following picture has the solution:

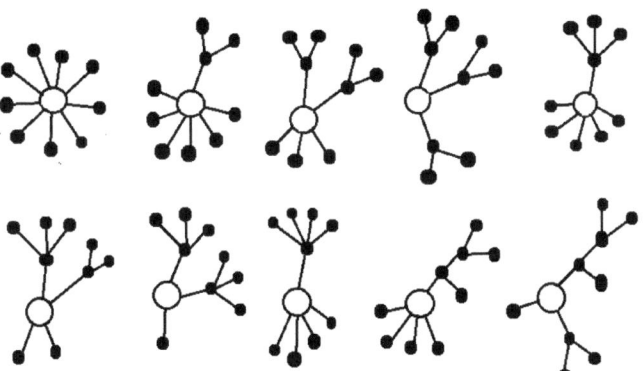

See those objects that consist of white and black vertices and the edges which connect the vertices to each other? They are trees, series-reduced free trees in this case. If you count the vertices in each tree, you will notice that every one of them has ten each. In other words, size 10 trees. The white vertices may seem a bit odd among the many black ones, but they do have a purpose: They represent the so-called root. A virtual root in this case, as free trees have no real roots in mathematics.

The root / virtual root is used in my ideas as a base of a tree when vertex edge algebra – the language of mathematical trees I have created – constructs the said tree. In the year 2020 I finally created the theorem that had already been present in my theories since 2014. Thanks to this theorem, vertex edge algebra can be used as a tree language. I call the theorem the theorem of sum forms, as vertex edge algebra constructs trees as sums of numbers, that now represent forms of these trees.

1.1Theorem of sum forms

" Any nonempty, finite tree can be represented as a combination consisting of three object types at most: ones 1, plus signs + and brackets ( ). "

Sort of a proof:

Any nonempty, finite tree consists of n vertices and n − 1 edges. The tree is connected so there are no isolated vertices (except for a single vertex, size one tree) and between any two arbitrarily chosen vertices there exists exactly one, unique path. Thus, we can choose a root/ virtual root, which is connected to every other vertex in the tree via a path. Let us express the root and every other vertex with a number one, edges with plus signs and a starting branch with brackets in a way that the root always starts the tree, and an internal vertex starts every branch. Other vertices are external vertices which are the last objects inside / outside branches and consecutive external vertices / branches at the same level of depth are connected to the root (outside brackets) / internal vertex (inside brackets), which starts that level of depth.

We construct a sum form, in which the root is connected via a path consisting of consecutive edges (+ signs) to any vertex in the tree. Examples:

1

Single root, size one tree.

$1 + 1$

One external vertex connected to the root. Size two tree.

$1 + 1 + 1$

Two external vertices are connected to the root. Size three tree.

$1 + (1 + 1)$

The root is connected to an internal vertex (the first number one inside the brackets), and this internal vertex is connected to an external vertex (the last number one inside the brackets). Size three tree.

$$1 + (1 + 1) + 1 + 1$$

The same as in the previous example, but the root is additionally connected to two new external vertices. Size five tree.

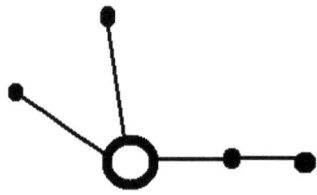

$$1 + 1 + 1 + 1 + 1 + 1$$

Five external vertices connected to the root. Size six tree.

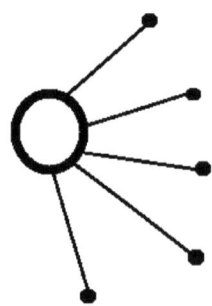

$$1 + \left(1 + \left(1 + (1 + 1)\right)\right) + 1$$

One external vertex is connected to the root, and also one branch, in which there are three consecutive internal vertices connected to each other in a line, and an external vertex connected to the last internal vertex. Size six tree.

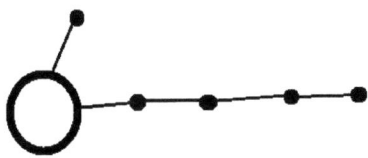

$$1 + (1 + 1 + 1) + 1$$

One external vertex and one internal vertex connected to the root. Additionally, the internal vertex is connected to two external vertices. Size five tree.

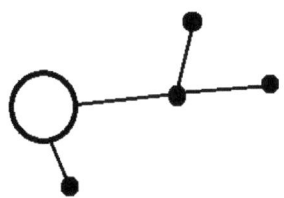

$$1 + (1 + (1 + (1 + 1 + 1) + 1) + 1)$$

In this and the further cases one should check out the picture, as writing and understanding the description may get quite laborious. Size eight tree.

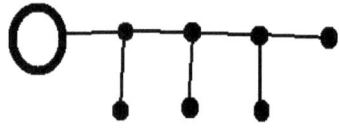

$$1 + \left(1 + (1 + 1 + 1 + 1)\right) + (1 + (1 + 1) + 1) + 1$$

Size 11 tree.

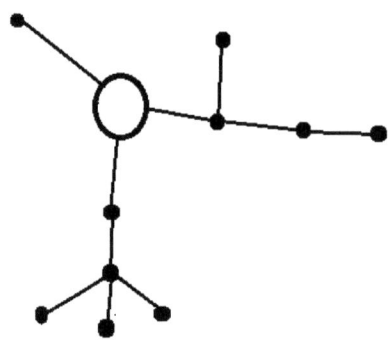

$1 + (1 + 1) + (1 + 1) + 1 + 1 + 1$

Size eight tree.

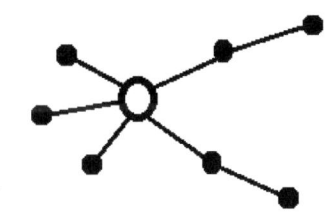

$1 + \left(1 + (1 + 1)\right) + (1 + 1 + 1) + 1 + 1$

Size nine tree.

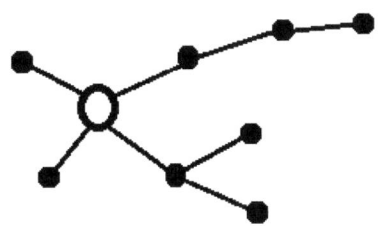

$$1 + (1 + 1 + 1) + (1 + 1) + 1$$

Size seven tree.

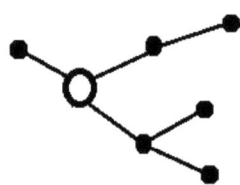

$$1 + (1 + (1 + 1) + 1) + (1 + 1) + 1$$

Size eight tree.

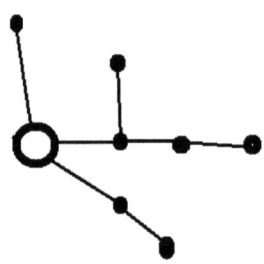

$$1 + (1 + (1 + 1 + 1 + 1) + 1 + 1) + (1 + 1 + 1) + 1$$

Size 12 tree.

Now as the theorem of sum forms has been presented, the collection of three main ideas I have created can now be examined. Vertex edge algebra is one of these ideas.

1.2Tget

Tget, tree generation and enumeration triplet, is a collection of three main ideas, which are the condition matrix, vertex edge algebra and tree generation algorithm. The algorithm also has two additional varieties: Branch form generator and tree enumeration algorithm.

1.2.1Condition matrix

Condition matrix $C_{v+x_i} = \{c_k\}_k$, is a collection of conditions c_k $(k \in \mathbb{N})$, where v is the sum of a tree's vertices (the types counted are defined by the condition v'), and x_i is the type of the condition matrix. The conditions together usually define the type of trees a forest – a collection of trees – has; the idea would be that only with knowing the properties of a tree could one find its forest and name in a possible database of condition matrices.

Tget focuses on four tree types:

1) Rooted trees

$$C_{n+x_{0.0.0}} = \{r_v[+], v = n, v' = rie, d_r \geq 1, d_i \geq 2, d_e = 1\} \quad (1)$$

2) Series-reduced rooted trees

$$C_{n+x_{0.0.1}} = \{r_v[+], v = n, v' = rie, d_r \neq 2, d_i \geq 3, d_e = 1\} \quad (2)$$

3) Free trees

$$C_{n+x_{1.0.0}} = \{r_v[-], v = n, v' = rie, d_r \geq 1, d_i \geq 2, d_e = 1\} \quad (3)$$

4) Series-reduced free trees

$$C_{n+x_{1.0.1}} = \{r_v[-], v = n, v' = rie, d_r \neq 2, d_i \geq 3, d_e = 1\} \quad (4)$$

$r_v[+]$ means that the root is treated as a root, so it is not a virtual root in this case, but a vertex with a label. $r_v[-]$, on the other hand, means that the root is virtual and only acts as a base for vertex edge algebraic operations. The same acting as a base also goes for the case $r_v[+]$, in addition to being as a root.

v is the sum of the vertices and v′ defines the types of vertices that are counted to create v − in this case, v′ = rie (root, internal, external vertex), so the vertex types which are counted are in this case the root, internal and external vertices. Thus, the number v is in this case the size of a tree in the forest. d_r is the number of vertices which are linked to the root, in other words it is the degree of the root. d_i is, on the other hand, the degree of the internal vertex and d_e is the degree of the external vertex. These have naturally positive nonzero integer values, except for the single root, size one tree where $d_r = 0$.

1.2.1.1Tree hierarchy

With the help of the condition matrix, collections of trees can be divided into four hierarchies: T_1 , T_2 , T_3 and T_4 .

T_4 is the collection of all the trees that exist in graph theory. It includes all the condition matrices with all the sizes $v \in \mathbb{N}$. Empty and infinite trees are also included, examples of condition matrices that produce empty and infinite trees:

$C_{0+x_{1.0.2}} = \{r_v[-], v = 0, v' = rie, d_r = 0, d_i \geq 2, d_e = 1\}$ produces one, empty tree, with the assumption that there is a free tree of size 0. It is also assumed that there is no rooted tree of size 0, as there is a rule. The rule states that there always must be a real root in a rooted tree, and if there are no vertices, then there is no real root either. This, however, does not apply to free trees, as they only have a virtual root, and the virtual root can be thought of defying the rule, at least if the rules of vertex edge algebra are bent.

$C_{8+x_{0.1.0}} = \{r_v[+], v = 8, v' = ri, d_r \geq 1, d_i \geq 2, d_e = 1\}$ produces infinite trees, as there can be an arbitrary number of external vertices (and thus for example branches of type ∞^x) and d_r and d_i are not restricted by an upper limit, thus allowing structures in which there are infinitely many external vertices connected to the root / internal vertex.

Versions that do not produce infinite trees:

$C_{8+x_{0.1.a}} = \{r_v[+], v = 8, v' = ri, d_r \in [1,7], d_i \in [2,6], d_e = 1\}$

$C_{8+x_{0.1.0.y}} = \{r_v[+], v = 8, v' = ri, d_r \geq 1, d_i \geq 2, d_e = 1, |t| = 16\}$

, with $|t|$ being the size of a tree (how many vertices in total).

$C_{10+x_{0.2.0}} = \{r_v[+], v = 10, v' = re, d_r \geq 1, d_i \geq 2, d_e = 1\}$ also produces infinite trees, as there can be an arbitrary number of internal vertices (and thus for example branches of type ∞^-) and $d_i \geq 2$ allows for infinitely long structures.

Versions that do not produce infinite trees:

$$C_{10+x_{0.2.1}} = \{r_v[+], v = 10, v' = re, d_r \neq 2, d_i \geq 3, d_e = 1\}$$

, with the assumption that the structures end with external vertices, even if they would be infinite.

$$C_{10+x_{0.2.0.d}} = \{r_v[+], v = 10, v' = re, d_r \geq 1, d_i \geq 2, d_e = 1, |t| = 14\}$$

T_3 is the collection of all the trees with all the possible sizes, defined by a specific condition matrix.

T_2 is the collection of all the trees with a specific size, defined by a specific condition matrix. In other words, T_2 is a forest.

T_1 is a tree in a specific collection of trees with a specific size, defined by a specific condition matrix.

1.2.2Vertex edge algebra

Trees can be expressed as sums of numbers, allowing the construction of a mathematical language called vertex edge algebra. This language is used by the tree generation algorithm and its varieties – branch form generator and tree enumeration algorithm. The symbols and rules for this language will now be presented.

1.2.2.1Symbols

1° is the root, which always starts the vertex edge algebraic representation of a tree – the sum form. There is exactly one root per one tree. For further clarification, one can make a notion 1^{o+}, if the sum form obeys the condition $r_v[+]$, and a notion 1^{o-} , if the sum form obeys the condition $r_v[-]$. If the condition is $r_v[+]$, then no two different sum forms have the same graphical representation, due to the real root. But, if the condition is $r_v[-]$, then the virtual root makes it possible that two different sum forms may have the same graphical representation.

1^* is an internal vertex and always starts a branch by being the first object inside the brackets. There is always one internal vertex per one bracket pair, although there can be several internal brackets inside brackets and thus several internal vertices, for example:

$$1^o + \left(1^* + \left(1^* + (1^* + 1^-) + (1^* + 1^- + 1^-)\right)\right) + 1^- .$$

1^- is an external vertex, which is in partition sum forms (sum forms represented by ordinary numbers) always the last object at every depth level, when every new internal bracket pair starts a new level. In normal symbolic sum forms, it does not matter whether the external vertex is the last object or not. Also, in partition sum forms, the branch/external vertex comes before the quantity coefficient.

$+$ is an edge and is always between two branches, two vertices, a vertex and a branch or a branch and a vertex. Outside brackets a branch or a vertex is connected with an edge to the root, whereas inside the brackets to the internal vertex which starts those brackets.

( ) are brackets, which always start a branch that has a size greater than one (two or more vertices). The first object inside the brackets is always an internal vertex.

$\cdot$ is a multiplication sign, which can make sum forms more compact. The notion $n^\sim \cdot m$ means, that there are m branches $n^\sim$ and $m - 1$ concealed + signs, $m \geq 1$. If $m = 0$, there are no concealed + signs, as the notation cannot even be opened (without the use of negative coefficients, which is a fringe area in vertex edge algebra). For example,

$$1^o + 3^x \cdot 2 + 1^- \cdot 3 = 1^o + 3^x + 3^x + 1^- + 1^- + 1^-.$$

n^x $(n \geq 3)$ is a branch, in which there is an internal vertex connected to $n - 1$ external vertices. In other words,

$$n^x = \left(1^* + 1^- \cdot (n - 1)\right).$$

Some examples are $3^x = (1^* + 1^- \cdot 2)$ and $4^x = (1^* + 1^- \cdot 3)$.

$(1^* \cdot n + \Sigma)$ is a notation, in which there are n consecutive internal vertices connected to each other in a line and an arbitrary collection Σ of branches and external vertices connected to the last internal vertex.

$$(1^* \cdot n + \Sigma) = \left(1^* + \left(1^* + (1^* + \ldots + (1^* + \Sigma)\ldots)\right)\right)$$

, in which there are n bracket pairs and n internal vertices.

Examples could be

$\Sigma = 3^x + 1^- \cdot 3$

$\Sigma = (1^* + 4^- + 2^-) + 5^-$.

Notice that

$(1^* \cdot 0 + \Sigma) = \Sigma$.

n^- ($n \geq 1$), on the other hand, is a branch in which there are $n - 1$ consecutive internal vertices connected to each other in a line and an external vertex connected to the last internal vertex. In other words,

$$n^- = \left(1^* + \left(1^* + (1^* + \ldots + (1^* + 1^-)\ldots)\right)\right)$$

, in which there are $n - 1$ bracket pairs and $n - 1$ internal vertices. Some examples are $2^- = (1^* + 1^-)$ and $3^- = (1^* + (1^* + 1^-))$.

Alternative notation:

$$n^- = (1^* \cdot (n - 1) + 1^-)$$

The branch n^- has the following property:

$$n^- = (1^* \cdot (n - a) + a^-), n \geq a \geq 0.$$

Example:

$$2^- = (1^* \cdot (2 - 0) + 0^-)$$
$$= (1^* \cdot 2 + 0^-)$$
$$= (1^* + (1^* + 0^-))$$
$$= \left(1^* + \left(1^* + 0^{()}\right)\right)$$
$$= (1^* + (1^*))$$
$$= (1^* + 1^-)$$

Some examples of graphical representations of trees and their symbolic sum forms:

The sum form in the picture: 1°

The sum form in the picture: $1^0 + 1^-$

The sum form in the picture: $1^0 + 1^- \cdot 2$

The sum form in the picture: $1^0 + (1^* + 1^-) = 1^0 + 2^-$

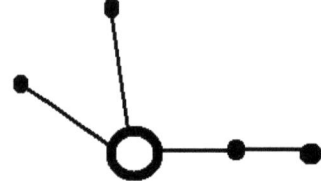

The sum form in the picture:

$$1^o + (1^* + 1^-) + 1^- \cdot 2 = 1^o + 2^- + 1^- \cdot 2$$

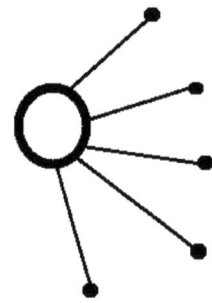

The sum form in the picture: $1^o + 1^- \cdot 5$

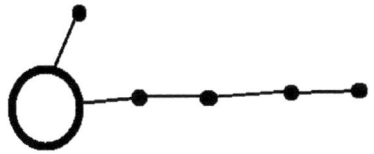

The sum form in the picture:

$$1^0 + (1^* \cdot 3 + 1^-) + 1^-$$
$$= 1^0 + (1^* + 3^-) + 1^-$$
$$= 1^0 + 4^- + 1^-$$

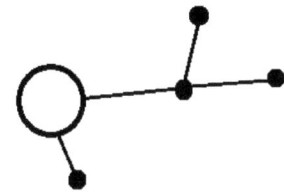

The sum form in the picture:

$$1^0 + (1^* + 1^- \cdot 2) + 1^- = 1^0 + 3^x + 1^-$$

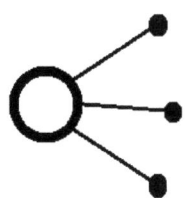

The sum form in the picture: $1^0 + 1^- \cdot 3$

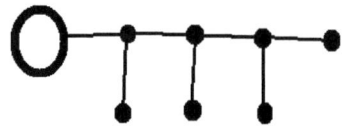

The sum form in the picture:

$$1^0 + (1^* + (1^* + 3^x + 1^-) + 1^-)$$

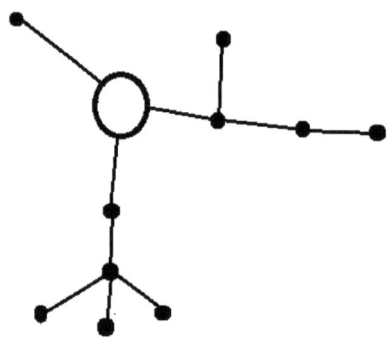

The sum form in the picture:

$$1^0 + (1^* + 4^x) + (1^* + 2^- + 1^-) + 1^-$$

$$= 1^0 + 5^{4Y} + 4^L + 1^-$$

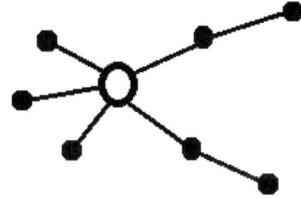

The sum form in the picture:

$$1^0 + (1^* + 1^-) \cdot 2 + 1^- \cdot 3$$

$$= 1^0 + 2^- \cdot 2 + 1^- \cdot 3$$

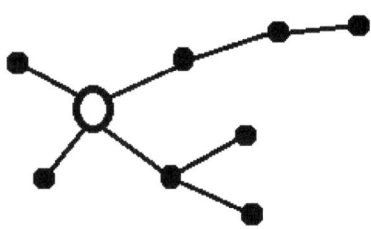

The sum form in the picture:

$$1^0 + \left(1^* + (1^* + 1^-)\right) + (1^* + 1^- + 1^-) + 1^- + 1^-$$

$$= 1^0 + 3^- + 3^x + 1^- \cdot 2$$

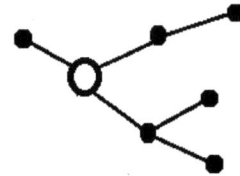

The sum form in the picture: $1^o + 3^x + 2^- + 1^-$

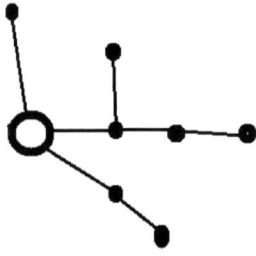

The sum form in the picture: $1^o + 4^L + 2^- + 1^-$

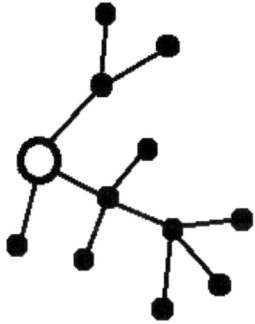

The sum form in the picture:

$$1^o + (1^* + 4^x + 1^- \cdot 2) + 3^x + 1^-.$$

n^{mL} ($n \geq 3$, $n \geq 2m + 1$, $m \geq 0$) is a branch, in which there is an internal vertex connected to m branches 2^- and also $n - 2m - 1$ external vertices. In other words,

$$n^{mL} = (1^* + 2^- \cdot m + 1^- \cdot (n - 2m - 1))$$

$$(2m + 1)^{mL} = (1^* + 2^- \cdot m)$$

$$n^{1L} = n^L$$

$$n^{0L} = n^x$$

$$3^{1L} = 3^-.$$

Some examples are $3^{1L} = (1^* + 2^-) = 3^-$, $4^L = (1^* + 2^- + 1^-)$, $5^{2L} = (1^* + 2^- \cdot 2)$, $7^{2L} = (1^* + 2^- \cdot 2 + 1^- \cdot 2)$, $7^{0L} = 7^x$.

$n^{mY}(n \geq m \geq 3)$ is a branch, in which there are $n - m$ consecutive internal vertices connected to each other in a line, and a branch m^x connected to the last internal vertex. In other words,

$$n^{mY} = \left(1^* + \left(1^* + (1^* + \ \dots + (1^* + m^x) \dots)\right)\right)$$

, in which there are $n - m$ bracket pairs and internal vertices. Some examples are $3^Y = 3^x, 4^Y = (1^* + 3^x)$, $5^{4Y} = (1^* + 4^x)$, $6^{3Y} = \left(1^* + \left(1^* + (1^* + 3^x)\right)\right)$.

$n^{3Y} = n^Y$

$m^{mY} = m^x$

Alternative notation:

$$n^{mY} = (1^* \cdot (n - m) + m^x)$$

$\doteqdot$ is a vertex sum equivalence notation and means that the sum forms on both sides of the notation have the same vertex sums v (v' defines what types are counted to create v).

For example, when v' = rie,

$$1^o + 1^- \cdot 2 \doteqdot 1^o + 2^-$$

, as $1 + 1 \cdot 2 = 1 + 2 = 3$.

It is good to know, that even though the vertex sums are the same, these sum forms are otherwise different (except for possible graphical equivalence). Generally speaking, a sum form is both identical and equal in vertex sum only with itself. In vertex edge algebra, when using the algorithms, one generates and enumerates all the different sum forms, and any identical ones are not allowed.

$(v_i)^{a_i o}_{rie\pm}$ is a closed form of a tree's sum form, in which v_i is the size of a tree and a_i is the tree's index in the sequence of all trees for the size v_i . Additional conditions are v' = rie and $r_v[+]$ for the + sign and $r_v[-]$ for the – sign. The index can be left out if it is not needed, thus using the form $(v_i)^{o}_{rie\pm}$.

$\{v_i\}^{\Sigma o}_{rie\pm}$ is a closed form of a forest's sum form, in which v_i is the size of a tree in the forest, with conditions v' = rie and $r_v[+]$ for the + sign and $r_v[-]$ for the – sign.

$\equiv$ is a forest equivalence notation, which is used between a forest's closed and opened sum forms. For example, for a rooted tree size 4 forest,

$$\{4\}^{\Sigma o}_{rie+} \equiv 1^o + 1^- \cdot 3 \doteq 1^o + 2^- + 1^- \doteq 1^o + 3^x \quad (5)$$
$$\doteq 1^o + 3^-$$

, where $\{4\}^{\Sigma o}_{rie+}$ is the forest's closed sum form (with conditions v' = rie and $r_v[+]$) and the right side of the notation $\equiv$ is the opened sum form of the forest.

$\cong$ is the graphical representation equivalence notation and is used between two sum forms, which have the same graphical representations. In the case $r_v[+]$ the sum form is graphically equivalent only with itself, the case $r_v[-]$ may be different. For example,

$$1^{o-} + 1^- \cdot 3 \cong 1^{o-} + 3^x \text{ and}$$

$$1^{o-} + 2^- \cdot 2 \cong 1^{o-} + 4^- \cong 1^{o-} + 3^- + 1^-.$$

$0^{()}$ is an empty branch, which has no real graphical representation. The size of an empty branch is zero, whereas the size of a branch $n^{\sim}$ is n.

$n^{a.-b.}$ is a notation, which is used in enumeration. It makes enumeration efficient and means, that the forms a. $-$ b. of the branch $n^{\sim}$ are gone through. For example, the case (5) can be rewritten as the following:

$$\{4\}_{rie+}^{\Sigma o} \equiv 1^o + 1^- \cdot 3 \doteq 1^o + 2^- + 1^- \quad (6)$$

$$\doteq 1^o + 3^{1.-2.} .$$

In rooted trees and free trees case, if there is a branch $n^{\sim}$ $(n \geq 3)$ with m forms, then the first form is n^x and the last form is n^-:

$$n^{1.} = n^x$$

and

$$n^{m.} = n^- .$$

If $m \geq 4$, then additionally

$$n^{2.} = n^L ,$$

$$n^{p.} = n^{(p-1)L} \ (n \geq 3, n \geq 2p - 1, p \geq 1)$$

and

$$n^{(m-1).} = n^Y.$$

$n\~(a, b)$ is also a notation, which is used in enumeration. It means, that the quantity coefficients a − b of the branch $n\~$ are gone through. The branch $n\~$ should only have one form, to simplify the case and ease the enumeration. Also, only the first branch which has only one form should be used in this notation, to further simplify the case. For example, in the series-reduced rooted and series-reduced free trees case, only the branch 3^x should be used with this notation, and the branch 4^x should be excluded.

For the rooted trees case v = 5,

$$\{5\}_{rie+}^{\Sigma o} \equiv 1^o + 1^- \cdot 4 \doteq 1^o + 2^- + 1^- \cdot 2$$

$$\doteq 1^o + 2^- \cdot 2$$

$$\doteq 1^o + 3^x + 1^-$$

$$\doteq 1^o + 3^- + 1^-$$

$$\doteq 1^o + 4^x$$

$$\doteq 1^o + (1^* + 2^- + 1^-)$$

$$\doteq 1^o + (1^* + 3^x)$$

$$\doteq 1^o + 4^\~ .$$

Or using the further notations as well:

$$\{5\}_{rie+}^{\Sigma o} \equiv 1^o + 1^- \cdot 4 \doteq 1^o + 2^- + 1^- \cdot 2$$

$$\doteq 1^o + 2^- \cdot 2$$

$$\doteq 1^o + 3^x + 1^-$$

$$\doteq 1^o + 3^- + 1^-$$

$$\doteq 1^o + 4^x$$

$$\doteq 1^o + 4^L$$

$$\doteq 1^o + 4^Y$$

$$\doteq 1^o + 4^- \, .$$

This can be simplified:

$$\{5\}_{rie+}^{\Sigma o} \equiv 1^o + 2^-(0,2) + 1^-(4,0) \quad (7)$$

$$\doteq 1^o + 3^{1.-2.} + 1^-$$

$$\doteq 1^o + 4^{1.-4.} \, .$$

The efficiency while using the enumeration tools grows quickly when the index v (in this case the tree size) and thus the rooted tree quantity r_v grow in size. This goes also for the quantity of series-reduced rooted trees s_v, quantity of free trees t_v and quantity of series-reduced free trees (homeomorphically irreducible trees) h_v.

Z is a number, which is used to express the sum form quantity for a certain row in enumeration. For example, for the case (7):

$$\{5\}_{rie+}^{\Sigma o} \equiv 1^o + 2^-(0,2) + 1^-(4,0) \ (Z = 3)$$

$$\doteq 1^o + 3^{1.-2.} + 1^-(Z = 2)$$
$$\doteq 1^o + 4^{1.-4.}(Z = 4)$$

, so $r_5 = 3 + 2 + 4 = 9$.

1.2.2.2Rules for symbolic sum forms

Let there be trees t_1, t_2 and t_3 with sum forms $(v_1)_{rie}^{a_1 o}$, $(v_2)_{rie}^{a_2 o}$ and $(v_3)_{rie}^{a_3 o}$. Thus

1) $t_i = t_i$, $(v_i)_{rie}^{a_i o} \doteq (v_i)_{rie}^{a_i o}$, $(v_i)_{rie}^{a_i o} = (v_i)_{rie}^{a_i o}$ and $(v_i)_{rie}^{a_i o} \cong (v_i)_{rie}^{a_i o}$, $i = 1,2,3$.

2) If $t_1 = t_2$, then these trees are the same.

3) If $(v_1)_{rie}^{a_1 o} \doteq (v_2)_{rie}^{a_2 o}$, then $v_1 = v_2$.

4) If $(v_1)_{rie}^{a_1 o} \cong (v_2)_{rie}^{a_2 o}$, then the graphical representations of these trees are the same.

5) If $t_1 = t_2$ and $t_2 = t_3$, then $t_1 = t_3$.

6) If $(v_1)_{rie}^{a_1 o} \doteq (v_2)_{rie}^{a_2 o}$ and $(v_2)_{rie}^{a_2 o} \doteq (v_3)_{rie}^{a_3 o}$, then $(v_1)_{rie}^{a_1 o} \doteq (v_3)_{rie}^{a_3 o}$.

7) If $(v_1)_{rie}^{a_1 o} \cong (v_2)_{rie}^{a_2 o}$ and $(v_2)_{rie}^{a_2 o} \cong (v_3)_{rie}^{a_3 o}$, then $(v_1)_{rie}^{a_1 o} \cong (v_3)_{rie}^{a_3 o}$.

8) If a branch / external vertex / an arbitrary collection of these has a quantity coefficient in a sum form, this quantity coefficient can come before or after the branch / external vertex / an arbitrary collection of these. However, there must also be a multiplication sign between these both. In the notation where an internal vertex has a quantity coefficient, the internal vertex always comes first. Also, in an empty tree, the root comes before the quantity coefficient (the coefficient is zero).

$$1^o + \Sigma_1 \cdot n + \Sigma_0 = 1^o + n \cdot \Sigma_1 + \Sigma_0 \, , n \geq 0$$

$$(1^* + \Sigma_1 \cdot n + \Sigma_0) = (1^* + n \cdot \Sigma_1 + \Sigma_0) \, , n \geq 0$$

, where Σ_0 and Σ_1 are arbitrary collections of branches and external vertices. For example,
$$\Sigma_0 = 1^-$$
$$\Sigma_1 = 3^x \cdot 3 + 2^-$$

$$1^o + \Sigma_1 \cdot 3 + \Sigma_0 = 1^o + (3^x \cdot 3 + 2^-) \cdot 3 + 1^-$$
$$= 1^o + 3^x \cdot 3 \cdot 3 + 2^- \cdot 3 + 1^-$$
$$= 1^o + 3 \cdot 3 \cdot 3^x + 3 \cdot 2^- + 1^-$$
$$= 1^o + 3 \cdot (3 \cdot 3^x + 2^-) + 1^-$$
$$= 1^o + 3 \cdot (3^x \cdot 3 + 2^-) + 1^- = 1^o + 3 \cdot \Sigma_1 + \Sigma_0$$

Additional examples:

$$1^o + 3 \cdot 3^x = 1^o + 3^x \cdot 3$$

$$1^o + (1 + 2 + 3) \cdot 6^- = 1^o + 6^- \cdot 1 + 6^- \cdot 2 + 6^- \cdot 3$$

$$1^o + 2 \cdot (2^- \cdot 4 + 5^L + 1337^x) + 4^Y + 1^- \cdot 5$$
$$= 1^o + (2^- \cdot 4 + 5^L + 1337^x) \cdot 2 + 4^Y + 1^- \cdot 5$$

$$(1^* + 2^- \cdot 7 + 3^x + 1^- + 1^-)$$
$$= (1^* + 7 \cdot 2^- + 3^x + 1^- + 1^-)$$

$$1^o + (1^* + 3^x) \cdot 2 = 1^o + 2 \cdot (1^* + 3^x).$$

9) If there are two or more branches / external vertices / arbitrary collections of these connected to the root or an internal vertex, then the arrangement of these (what comes first, what is in the middle and what comes last in the sum form etc.) in that specific vertex does not matter. However, the root always starts the sum form, and the internal vertex always starts a branch (size larger than one) with brackets. In a symbolic sum form, it does

not matter whether the external vertex is the last object or not.

$$1^o + \Sigma_1 + \Sigma_2 = 1^o + \Sigma_2 + \Sigma_1$$

$$(1^* + \Sigma_1 + \Sigma_2) = (1^* + \Sigma_2 + \Sigma_1)$$

, where Σ_1 and Σ_2 are arbitrary collections of branches and external vertices. For example,

$$\Sigma_1 = 66^x$$
$$\Sigma_2 = 8^- \cdot 5 + (1^* + (1^* \cdot 5 + 3 \cdot 4^L + 5^Y) + 1^-)$$

Examples:

$$1^o + 2^- + 3^x + 4^L + 5^Y$$
$$= 1^o + 4^L + 5^Y + 2^- + 3^x$$

$$1^o + 2^- + 3^- + 1^- = 1^o + 3^- + 2^- + 1^-$$

$$1^o + (1^* + 4^x + 2^- \cdot 3) + 3^x = 1^o + 3^x +$$
$$(1^* + 3 \cdot 2^- + 4^x) .$$

10) There is always one root per one sum form and one internal vertex per one bracket pair in

a way that the internal vertex starts those brackets. However, there can be several brackets and internal brackets and thus several internal vertices at depth levels that are getting deeper and deeper.

11) + sign, an edge, is only used between two branches, two vertices, a branch and a vertex or a vertex and a branch, when considering fully opened sum forms (no quantity coefficients or unnecessary brackets, brackets of opened branches are necessary). Outside brackets the + sign connects the object (a branch or a vertex) on its right side to the root, which starts the sum form; inside the brackets the + sign connects the object (a branch or a vertex) on its right side to the internal vertex, which starts those brackets. What comes to addition, the + sign can also be used between two quantity coefficients of the same branch / external vertex, if these coefficients are inside brackets.

12) · sign is only used between a quantity coefficient and a branch, a branch and a

quantity coefficient, a quantity coefficient and an external vertex, an external vertex / an internal vertex and a quantity coefficient, a quantity coefficient and an arbitrary collection of branches and external vertices, or an arbitrary collection of branches and external vertices and a quantity coefficient. However, a quantity coefficient can also be multiplied with another quantity coefficient, if these coefficients are for the same branch / external vertex / collection.

$$\Sigma \cdot n = \sum_{k=1}^{n} \Sigma, n \geq 1$$
$$\Sigma \cdot 0 = 0^{()}$$

, where Σ is an arbitrary collection of branches and external vertices.

Examples:

$$1^o + 4^x \cdot 4 = 1^o + 4^x + 4^x + 4^x + 4^x$$

$$1^o + (1^- + 2^- + 3^-) \cdot 0 = 1^o + 0^{()} = 1^o$$

13) For rooted trees and free trees case, if there is a branch $n^{\sim}$ ($n \geq 3$) with $m \geq 2$ forms, then the form n^{*} is the first and the form n^{-} is the last form.

14) If an empty branch is connected to the root or an internal vertex, the case is treated as if no branch was connected to them in the first place. Even if there is a quantity coefficient or not.

$$1^{o} + \Sigma + 0^{()} \cdot n = 1^{o} + \Sigma \, , n \geq 0$$

$$\left(1^{*} + \Sigma + 0^{()} \cdot n\right) = (1^{*} + \Sigma) \, , n \geq 0$$

$$0^{()} \cdot n = 0^{()} \, , n \geq 0$$

, where Σ is an arbitrary collection of branches and external vertices.

Examples:

$$1^{o} + 2^{-} + 3^{x} + 0^{()} = 1^{o} + 2^{-} + 3^{x}$$

$$1^{o} + 3^{L} \cdot 5 + 0^{()} \cdot 4 = 1^{o} + 3^{L} \cdot 5 \, .$$

15) If a branch / an external vertex / an arbitrary collection of these, has a quantity coefficient of one, then this quantity coefficient does not have to be shown. If, however, the branch / external vertex / arbitrary collection of these, has a quantity coefficient of zero, then this object does not exist in the sum form. In this case, neither the branch / external vertex / arbitrary collection of these, nor the quantity coefficient, must be shown. But if one must be shown, the other must also be.

$$\Sigma \cdot 1 = \Sigma$$

$$\Sigma \cdot 0 = 0^{()}$$

, where Σ is an arbitrary collection of branches and external vertices.

Examples:

$$1^o + 23^x \cdot 5 + 1^- \cdot 1 = 1^o + 23^x \cdot 5 + 1^-$$

$$1^o + 6^- \cdot 0 + (1^* + 1^-) = 1^o + (1^* + 1^-).$$

$$(1^* + (2^- + 2 \cdot 4^x) \cdot 0) = (1^* + 0^{()}) = (1^*) = 1^-$$

16) If there are k_1 branches $n_{\tilde{1}}$ (number of forms Z_1 per one branch), k_2 branches $n_{\tilde{2}}$ (number of forms Z_2 per one branch), ... , k_m branches $n_{\tilde{m}}$ (number of forms Z_m per one branch), then the whole quantity Z of sum forms on the row is

$$Z = \prod_{i=1}^{m} \binom{Z_i + k_i - 1}{k_i} \qquad (8)$$

Example:

The row $1^0 + 4^{1.-4.} \cdot 2 + 3^{1.-2.} \cdot 3 + 2^{-} \cdot 10 + 1^{-}$ has the following number Z:

$$Z = \binom{4+2-1}{2}\binom{2+3-1}{3}\binom{1+10-1}{10}\binom{1+1-1}{1}$$

$$= \binom{5}{2}\binom{4}{3}\binom{10}{10}\binom{1}{1}$$

$$= \frac{5 \cdot 4 \cdot (3)!}{2!\,(5-2)!} \cdot \frac{4 \cdot (3)!}{3!\,(4-3)!} \cdot \frac{10!}{10!\,(0)!} \cdot \frac{1!}{1!\,(0)!}$$

$$= \frac{20 \cdot (3)!}{2 \cdot (3)!} \cdot \frac{4 \cdot (3)!}{1!\,(3)!} \cdot \frac{10!}{10!} \cdot \frac{1}{1} = \frac{20}{2} \cdot \frac{4}{1} \cdot 1 \cdot 1 = 10 \cdot 4$$

$$= 40.$$

1.2.2.3 Known graphical identities

1) $1^{o-} + 1^{-} \cdot n \cong 1^{o-} + n^{x}$; $n \geq 3$

2) $1^{o-} + n^{-} \cong 1^{o-} + (n-a)^{-} + (a)^{-}$

   ; $0 \leq a \leq n$

3) $1^{o-} + (1^{*} + \Sigma_1) + \Sigma_2 \cong 1^{o-} + (1^{*} + \Sigma_2) + \Sigma_1$

; where Σ_1, Σ_2 are arbitrary collections of branches and external vertices. Examples could be:

$\Sigma_1 = 3^{x} + 2^{-} \cdot 2 + 1^{-}$

$\Sigma_2 = 15^{-} + (1^{*} + 5^{x} + (1^{*} + 1^{-} \cdot 3) + 1^{-}) + 2^{-}$

4) $1^{o-} + (1^{*} + n^{\sim}) + m^{\sim} \cong 1^{o-} + (1^{*} + m^{\sim}) + n^{\sim}$

5) $1^{o-} + n^{mL} \cong 1^{o-} + 2^{-} \cdot m + 1^{-} \cdot (n - 2m)$; $n \geq 2m + 1$, $n \geq 3$, $m \geq 0$

6) $1^{o-} + (2n+1)^{nL} \cong 1^{o-} + 2^- \cdot n + 1^-$;
   $n \geq 1$

7) $1^{o-} + n^L \cong 1^{o-} + (n-1)^x + 1^-; n \geq 4$

8) $1^{o-} + n^{mL} + 1^- \cdot k \cong 1^{o-} + (k+1)^x +$
   $2^- \cdot m + 1^- \cdot (n - 2m - 1)$, $n \geq 2m +$
   $1, n \geq 3, m \geq 0, k \geq 2$

9) $1^{o-} + n^L \cong 1^{o-} + n^{(n-1)Y}, n \geq 4$

10) $1^{o-} + n^Y \cong 1^{o-} + (1^* + (n-2)^- + 1^-);$
    $n \geq 3$

11) $1^{o-} + n^Y + m^- \cong 1^{o-} + (n+m)^Y$
    $; n \geq 3, m \geq 0$

12) $1^{o-} + n^{mY} \cong 1^{o-} + (n - m + 1)^- + 1^- \cdot$
    $(m-1); n \geq m \geq 3$

13) $1^{o-} + m^x + n^- \cong 1^{o-} + (n+m)^{mY}$
    $; m \geq 3; n \geq 0$

14) $1^{o-} + n^- + 1^- \cdot 2 \cong 1^{o-} + (n+2)^Y$
    $; n \geq 1$

15) $1^{o-} + 2^- \cdot n + 1^- \cong 1^{o-} + \left(1^* + (1^* + 2^- \cdot (n-1) + 1^-)\right); n \geq 1$

16) $1^{o-} + n^- + \Sigma \cong 1^{o-} + (1^* \cdot n + \Sigma); n \geq 0$

The case 4) is a special case of the case 3). In the case 4) the branches $n^\sim, m^\sim$ can have any of their forms, and the cases $n^\sim = m^\sim$ and $n = m$ are possible (in the case $n = m$ the branches can still have different forms). Σ is an arbitrary collection of branches and external vertices. Examples:

$n^\sim = 3^x$

$m^\sim = (1^* + (1^* + 1^-) + 2^- + 4^L + 1^- \cdot 4)$

$\Sigma = (1^* + (1^* \cdot 3 + 5^Y + 2^-) + 3^x \cdot 2) + 1^-$

1.2.2.4Known facts

1) $1^o + \Sigma_1 \cdot n + \Sigma_0 = 1^o + n \cdot \Sigma_1 + \Sigma_0$, $n \geq 0$

2) $(1^* + \Sigma_1 \cdot n + \Sigma_0) = (1^* + n \cdot \Sigma_1 + \Sigma_0)$

   , $n \geq 0$

3) $1^o + \Sigma_1 + \Sigma_2 = 1^o + \Sigma_2 + \Sigma_1$

4) $(1^* + \Sigma_1 + \Sigma_2) = (1^* + \Sigma_2 + \Sigma_1)$

5) $1^o + \Sigma + 0^{()} \cdot n = 1^o + \Sigma$, $n \geq 0$

6) $(1^* + \Sigma + 0^{()} \cdot n) = (1^* + \Sigma)$, $n \geq 0$

7) $n^x = \left(1^* + 1^- \cdot (n-1)\right)$; $n \geq 3$

8) $n^- = (1^* \cdot (n-1) + 1^-)$; $n \geq 1$

9) $n^- = (1^* \cdot (n-a) + a^-)$, $n \geq a \geq 0$

10) $n^{mL} = (1^* + 2^- \cdot m + 1^- \cdot (n - 2m - 1))$;
    $n \geq 3, n \geq 2m + 1, m \geq 0$

11) $(2m + 1)^{mL} = (1^* + 2^- \cdot m)$; $m \geq 1$

12) $n^{1L} = n^L$; $n \geq 3$

13) $n^{0L} = n^x$; $n \geq 3$

14) $3^L = 3^-$

15) $n^{mY} = (1^* \cdot (n - m) + m^x)$

   ; $n \geq m \geq 3$

16) $m^{mY} = m^x$; $m \geq 3$

17) $n^{3Y} = n^Y$; $n \geq 3$

18) $(1^* \cdot 0 + \Sigma) = \Sigma$

19) $(1^* \cdot 1 + \Sigma) = (1^* + \Sigma)$

20) $(1^* + 0^{()}) = (1^*) = 1^-$

21) $(1^* + 0^{()}) \cdot n = 1^- \cdot n$; $n \geq 0$

22) $0^- = 0^x = 0^{()} = 0^{mL} = 0^{mY}$

23) $\Sigma_1 \cdot 1 = \Sigma_1$

24) $\Sigma_1 \cdot 0 = 0^{()}$

25) $\Sigma \cdot n = \sum_{k=1}^{n} \Sigma , n \geq 1$

$\Sigma_0, \Sigma_1 , \Sigma_2$ and Σ are arbitrary collections of branches and external vertices. For example:

$\Sigma_0 = 2^- \cdot 2$
$\Sigma_1 = 1^- \cdot 2 + 3^x$
$\Sigma_2 = (1^* + 5^x + 4^L + 1^-) + 0^{()}$
$\Sigma = 100^{12L}$

1.2.2.5Number of graphically identical forms

All the trees have a size m. N is the number of graphically identical forms, when the sum form itself is included.

Sum form	N
1^{o-}	1
$1^{o-} + 1^-$	1
$1^{o-} + 1^- \cdot n$, $n \geq 2$	2
$1^{o-} + n^-$, $n \geq 0$	$\left\lceil \dfrac{m}{2} \right\rceil$
$1^{o-} + n^x + 1^- \cdot (n-1)$, $n \geq 3$	2
$1^{o-} + \sum_{k=1}^{n} k^-$, $n \geq 3$	m
$1^{o-} + 2^- \cdot n + 1^- \cdot k$, $n \geq 1, k \geq 2$	4
$1^{o-} + n^Y$, $n \geq 4$	$m - 1$

1.2.2.6Bending the rules of vertex edge algebra

1. The empty tree

$$(0)^o_{rie-} = 1^o \cdot 0 + (1^* \cdot 0 + 1^- \cdot 0)$$
$$= 1^o \cdot 0 + 0^{()}$$
$$= 1^o \cdot 0$$

2. Infinite trees

$$(\infty)^o_{rie\pm} = 1^o + 1^- \cdot \infty$$

$$(\infty)^o_{rie\pm} = 1^o + \infty^x$$

$$(\infty)^o_{rie\pm} = 1^o + 2^- + 3^x + 4^x + 5^x + \cdots$$

$$(\infty)^o_{rie\pm} = 1^o + 1 \cdot (1^* + 2 \cdot (1^* + \cdots + n \cdot (1^* + 1^-) \ldots)), n \to \infty$$

$$(\infty)^o_{rie\pm} = 1^o + (1 + 2 + 3 + \cdots)^-$$

$$(\infty)^o_{rie\pm} = \lim_{n \to \infty} (1^o + 2^- + 3^- + \cdots + n^-)$$

$$(\infty)^o_{rie\pm} = 1^o + \sum_{k=3}^{\infty}(k^L) + 1^-$$

3. Alternative methods to represent trees

$$(56)^o_{rie\pm} = 1^o + \sum_{m=2}^{10}(n^-) + 1^-$$

$$(22)^o_{rie\pm} = 1^o + 1^- + 2^- + \cdots + 6^-$$

$$(37)^o_{rie\pm} = 1^o + (1 \cdot 2 \cdot 3) \cdot (1^- + 2^- + 3^x)$$

$$(57)^o_{rie\pm} = 1^o + 4! \cdot 1^- + (2^5)^x$$

4. Coefficients which are not positive integers or zero

$$(2)^o_{rie\pm} = 1^o + 1^- \cdot \frac{1}{2} + 1^- \cdot \frac{1}{2}$$

$$(3)^o_{rie\pm} = 1^o + \frac{1}{2} \cdot 2^- + \frac{1}{3} \cdot 2^- + \frac{1}{6} \cdot 2^-$$

$$(9)^o_{rie\pm} = 1^o + 2 \cdot 4^x + (-2) \cdot (4^x) + 8^{4Y}$$

1.2.3 Tree generation algorithm and its variations

The tree generation algorithm is an algorithm, which uses vertex edge algebra as its language. Another option would be to use partition language (sums of numbers). The algorithm can be used without computers to generate and enumerate trees, that may give an opportunity compared to other machine-based algorithms. By analyzing the outputs of the tree generation algorithm and its variations one can get deep insights into how an efficient and working tree generation and enumeration system works.

The tree generation algorithm's process for rooted trees is the following:

1.2.3.1Tree generation algorithm

1) Let $v = n$.

2) Set the root as a static base, with removing number one (1) from the number n. The root is the first number in the sum forms and does not take part in the partitions. Sum forms contain both the bases and the partitions.

3) Solve the partition equation

$$\sum_{i=1}^{m}(x_i) = n - 1 \, , \; m \leq n - 1 \, , \; x_i \in [1, n - 1]$$
(9)

, with inserting each partition into its own column with the exception that the sum forms that have the same size largest branches go to the same columns. The order in the columns is from up to down, and the sum forms with the largest column specific branch $n\tilde{}$ are in the column C_n , when the column's C_1 largest sum form object is an external vertex.

4) Repeat the part 3) with setting the internal vertices, the first number ones (1) inside brackets (branches), as static bases, and then solving the partition equation. This happens with removing number one (1) from the size of the branch $p\tilde{}$ (size is greater than two, nothing happens to size two branch), and then solving the partition equation

$$\sum_{i=1}^{m}(x_i) = p - 1 \, , \; m \leq p - 1 \, , \; x_i \in [1, p - 1]$$

Remember that the number ones set as static bases do not take part in the partitions. The whole sum forms that contain these partitions, are put into the same columns where the original sum forms were. If there are branches with different sizes, then the forms of the largest ones are gone through first. If there are branches with same sizes, then it does not matter which forms of a particular branch are gone through first, if the following rule is considered.

Generally, when there are k branches $n^{\sim}$ (m forms per branch), the corresponding combination codes

$$C_{1;2;3;\ldots;k} = (a_1, a_2, a_3, \ldots, a_k) \quad ; 1 \leq a_j \leq m$$

, follow the following rule:

$$a_1 \leq a_2 \leq a_3 \leq \cdots \leq a_k .$$

$C_{1;2;3;\ldots;k}$ corresponds to a group of branches

$$n^{a_1}, n^{a_2}, n^{a_3}, \ldots, n^{a_k}.$$

If put into a sum form, they would form the following form:

$$1^o + n^{a_k} + n^{a_{k-1}} + \cdots + n^{a_3} + n^{a_2} + n^{a_1}.$$

Examples:

Let there be a sum form, which has branches 3~ (two forms), 3~ (two forms) and 4~ (four forms). Then the combinations are the following:

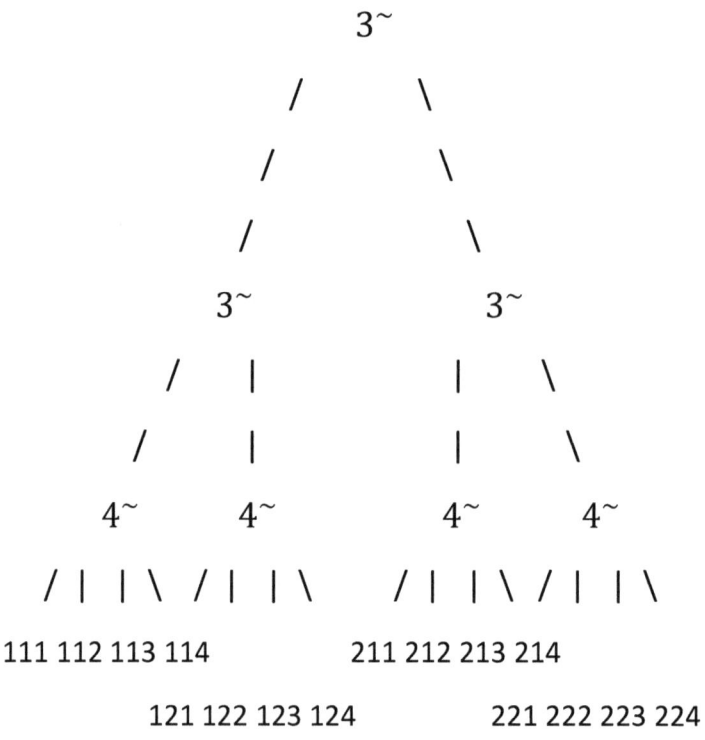

Thus, if the sum form has a form $1^o + 4^x + 3^x \cdot 2$ (code (1,1,1)), then the sum forms for the tree generation algorithm case would be the following, when the combination codes (2,1,X) are discarded due to similarity:

Code (1,1,1): $1^o + 4^x + 3^x \cdot 2$

Code (1,1,2): $1^o + 4^{2.} + 3^x \cdot 2$

Code (1,1,3): $1^o + 4^{3.} + 3^x \cdot 2$

Code (1,1,4): $1^o + 4^{4.} + 3^x \cdot 2$

Code (1,2,1): $1^o + 4^{1.} + 3^- + 3^x$

Code (1,2,2): $1^o + 4^{2.} + 3^- + 3^x$

Code (1,2,3): $1^o + 4^{3.} + 3^- + 3^x$

Code (1,2,4): $1^o + 4^{4.} + 3^- + 3^x$

Code (2,2,1): $1^o + 4^{1.} + 3^- \cdot 2$

Code (2,2,2): $1^o + 4^{2.} + 3^- \cdot 2$

Code (2,2,3): $1^o + 4^{3.} + 3^- \cdot 2$

Code (2,2,4): $1^o + 4^{4.} + 3^- \cdot 2$

$$3^{1.} = 3^x$$
$$3^{2.} = 3^-$$
$$4^{1.} = 4^x$$
$$4^{2.} = (1^* + 2^- + 1^-)$$
$$4^{3.} = (1^* + 3^x)$$
$$4^{4.} = 4^- .$$

Let there be another sum form, which has branches 4~
(four forms) and 4~ (four forms). Then the combinations
are the following:

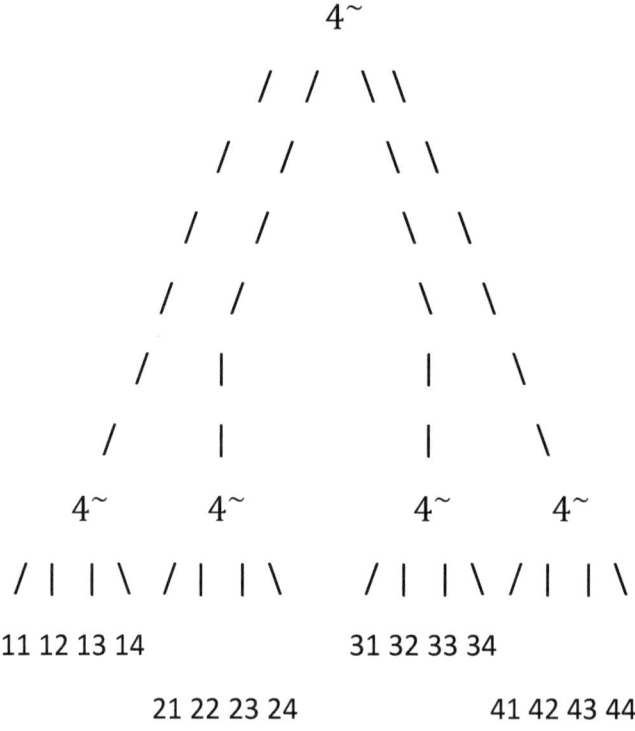

Thus, if the sum form has a form $1^o + 4^x \cdot 2$ (code (1,1)), then the sum forms for the tree generation algorithm case would be the following, when the combination codes (2,1), (3,1), (3,2), (4,1), (4,2) and (4,3) are discarded due to similarity:

Code (1,1): $1^o + 4^{1.} \cdot 2$

Code (1,2): $1^o + 4^{2.} + 4^{1.}$

Code (1,3): $1^o + 4^{3.} + 4^{1.}$

Code (1,4): $1^o + 4^{4.} + 4^{1.}$

Code (2,2): $1^o + 4^{2.} \cdot 2$

Code (2,3): $1^o + 4^{3.} + 4^{2.}$

Code (2,4): $1^o + 4^{4.} + 4^{2.}$

Code (3,3): $1^o + 4^{3.} \cdot 2$

Code (3,4): $1^o + 4^{4.} + 4^{3.}$

Code (4,4): $1^o + 4^{4.} \cdot 2$

$$3^{1.} = 3^x$$
$$3^{2.} = 3^-$$
$$4^{1.} = 4^x$$
$$4^{2.} = (1^* + 2^- + 1^-)$$
$$4^{3.} = (1^* + 3^x)$$
$$4^{4.} = 4^- \ .$$

Last example: sum form, which has branches $3^\sim$(two forms), $3^\sim$ (two forms) and $3^\sim$ (two forms). Then the combinations are the following:

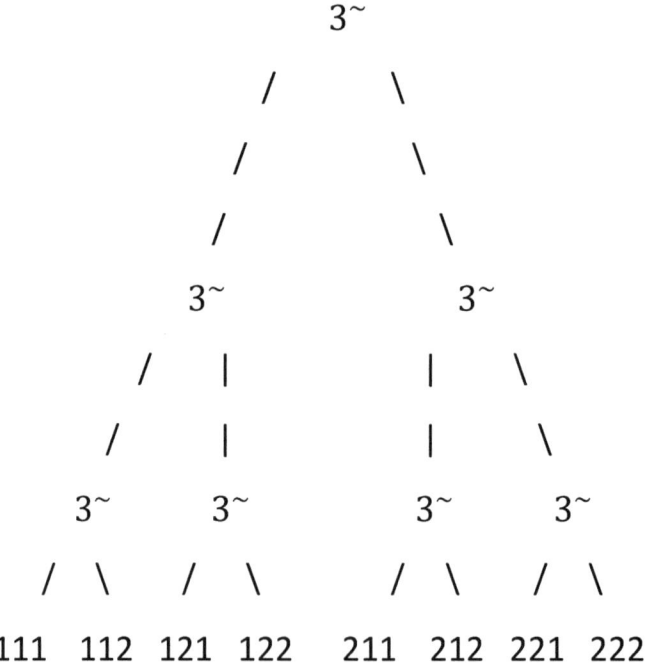

Thus, if the sum form has a form $1^o + 3^x \cdot 3$ (code (1,1,1)), then the sum forms for the tree generation algorithm case would be the following, when the combination codes (1,2,1), (2,1,1), (2,1,2) and (2,2,1) are discarded due to similarity:

Code (1,1,1): $1^o + 3^x \cdot 3$

Code (1,1,2): $1^o + 3^- + 3^x \cdot 2$

Code (1,2,2): $1^o + 3^- \cdot 2 + 3^x$

Code (2,2,2): $1^o + 3^- \cdot 3$

$$3^{1.} = 3^x$$
$$3^{2.} = 3^-$$

5) Repeat the part 4) until every partition has been gone through lexicographically and no further partitions are possible. Link every partition with inserting an equality sign between two consecutive partitions. Also, link the first partition with the number n (forest).

6) Change all the symbols in the following way:

6.1) The first number n, which has all the partitions, is the closed forest with a tree size n. The first equality sign is the symbol $\equiv$.

6.2) The first number one (1) outside brackets is the root, the first number one inside brackets (every starting bracket pair) is an internal vertex and every other one (inside and outside brackets) is an external vertex.

6.3) Every other equality sign is a vertex sum equivalence sign. The numbers on the right side of the multiplication sign $\cdot$ are quantity coefficients and the numbers on the left side of the multiplication sign are branches / external vertices.

6.4) The number two is a branch 2^- and the number k (k $\geq$ 3) is a branch k^x. The objects
$$(1+1), (1+(1+1)) = (1+2), \left(1+(1+(1+1))\right) = (1+(1+2)), \dots ,$$
can be compactified with the notions $2^-, 3^-, 4^-, \dots$
.

1.2.3.2Examples:

i.   Let there be a rooted tree forest with tree size $v = 5$. The tree generation algorithm works as in the procedure. The partitions are already linked with equality signs in this and the following examples.

1) $v = 5$

2) $5 = 1 + 4 = 1 + 4 = 1 + 4 = 1 + 4$
$$= 1 + 4$$

3) $5 = 1 + 1 \cdot 4 = 1 + 2 + 1 \cdot 2$
$$= 1 + 2 \cdot 2$$
$$= 1 + 3 + 1$$

$$= 1 + 4 \, .$$

4) $5 = 1 + 1 \cdot 4 = 1 + 2 + 1 \cdot 2$
$$= 1 + 2 \cdot 2$$

$= 1 + (1 + 1 \cdot 2) + 1$

$= 1 + (1 + 2) + 1$

$= 1 + (1 + 1 \cdot 3)$

$= 1 + (1 + 2 + 1)$

$= 1 + (1 + 3) \,.$

5) $5 = 1 + 1 \cdot 4 = 1 + 2 + 1 \cdot 2$
$$= 1 + 2 \cdot 2$$

$= 1 + (1 + 1 \cdot 2) + 1$

$= 1 + (1 + 2) + 1$

$= 1 + (1 + 1 \cdot 3)$

$= 1 + (1 + 2 + 1)$

$= 1 + \big(1 + (1 + 1 \cdot 2)\big)$

$= 1 + \big(1 + (1 + 2)\big) \,.$

6) $\{5\}_{rie+}^{\Sigma o} \equiv 1^o + 1^- \cdot 4$

$\doteq 1^o + 2^- + 1^- \cdot 2$

$\doteq 1^o + 2^- \cdot 2$

$\doteq 1^o + 3^x + 1^-$

$\doteq 1^o + 3^- + 1^-$

$\doteq 1^o + 4^x$

$\doteq 1^o + (1^* + 2^- + 1^-)$

$\doteq 1^o + (1^* + 3^x)$

$\doteq 1^o + 4^- \ .$

ii. Another rooted trees example, with tree size v = 6. The tree generation algorithm works as in the process.

1) $v = 6$

2) $6 = 1 + 5 = 1 + 5$
$$= 1 + 5$$
$$= 1 + 5$$
$$= 1 + 5$$
$$= 1 + 5$$
$$= 1 + 5$$

3) $6 = 1 + 1 \cdot 5 = 1 + 2 + 1 \cdot 3$
$$= 1 + 2 \cdot 2 + 1$$
$$= 1 + 3 + 1 \cdot 2$$

$$= 1 + 3 + 2$$

$$= 1 + 4 + 1$$

$$= 1 + 5 \, .$$

4) and 5) $6 = 1 + 1 \cdot 5 = 1 + 2 + 1 \cdot 3$
$$= 1 + 2 \cdot 2 + 1$$

$$= 1 + (1 + 1 \cdot 2) + 1 \cdot 2$$
$$= 1 + (1 + 2) + 1 \cdot 2$$
$$= 1 + (1 + 1 \cdot 2) + 2$$
$$= 1 + (1 + 2) + 2$$

$$= 1 + (1 + 1 \cdot 3) + 1$$
$$= 1 + (1 + 2 + 1) + 1$$
$$= 1 + \big(1 + (1 + 1 \cdot 2)\big) + 1$$
$$= 1 + \big(1 + (1 + 2)\big) + 1$$

$$= 1 + (1 + 1 \cdot 4)$$

$$= 1 + (1 + 2 + 1 \cdot 2)$$

$$= 1 + (1 + 2 \cdot 2)$$

$$= 1 + (1 + (1 + 1 \cdot 2) + 1)$$

$$= 1 + (1 + (1 + 2) + 1)$$

$$= 1 + \left(1 + (1 + 1 \cdot 3)\right)$$

$$= 1 + \left(1 + (1 + 2 + 1)\right)$$

$$= 1 + \left(1 + \left(1 + (1 + 1 \cdot 2)\right)\right)$$

$$= 1 + \left(1 + \left(1 + (1 + 2)\right)\right).$$

6) $\{6\}_{rie+}^{\Sigma o} \equiv 1^o + 1^- \cdot 5 \doteq 1^o + 2^- + 1^- \cdot 3$

$$\doteq 1^o + 2^- \cdot 2 + 1^-$$

$\doteq 1^o + 3^x + 1^- \cdot 2$

$\doteq 1^o + 3^- + 1^- \cdot 2$

$\doteq 1^o + 3^x + 2^-$

$\doteq 1^o + 3^- + 2^-$

$$\doteq 1^o + 4^x + 1^-$$

$$\doteq 1^o + (1^* + 2^- + 1^-) + 1^-$$

$$\doteq 1^o + (1^* + 3^x) + 1^-$$

$$\doteq 1^o + 4^- + 1^-$$

$$\doteq 1^o + 5^x$$

$$\doteq 1^o + (1^* + 2^- + 1^- \cdot 2)$$

$$\doteq 1^o + (1^* + 2^- \cdot 2)$$

$$\doteq 1^o + (1^* + 3^x + 1^-)$$

$$\doteq 1^o + (1^* + 3^- + 1^-)$$

$$\doteq 1^o + (1^* + 4^x)$$

$$\doteq 1^o + \left(1^* + (1^* + 2^- + 1^-)\right)$$

$$\doteq 1^o + \left(1^* + (1^* + 3^x)\right)$$

$$\doteq 1^o + 5^- \,.$$

iii. Series-reduced rooted trees example. The tree generation algorithm works now differently:

a) The type n^-, $n \geq 2$, branches are forbidden.

b) Outside brackets the number of objects (branch, external vertex) connected to the root must not be equal to two.

c) In a branch there must be at least two objects (branch, external vertex) connected to the internal vertex inside brackets. Thus, there must be three or more objects connected to the internal vertex in total, when the base connection point for the branch is taken into account. This rule applies for every internal vertex and bracket pair.

1) $v = 6$

2) $6 = 1 + 5 = 1 + 5 = 1 + 5$

3) , 4) and 5)

$$6 = 1 + 1 \cdot 5 = 1 + 3 + 1 \cdot 2$$

$$= 1 + (1 + 1 \cdot 4)$$
$$= 1 + (1 + 3 + 1) \, .$$

6) $\{6\}_{rie+}^{\Sigma o} \equiv 1^o + 1^- \cdot 5 \doteq 1^o + 3^x + 1^- \cdot 2$

$\doteq 1^o + 5^x$

$\doteq 1^o + (1^* + 3^x + 1^-) \, .$

iv. Free trees example. The same rules as for the rooted trees, but with the following exceptions:

a) Let $v = i + j = n > 5$; $i \leq j$; i and j are positive nonzero integers. The cases $v = 1 - 5$ will be shown after the parts a), b) and c).

b) The columns j are discarded.

c) In the case $v = n = 2p = i + j$, in the column $i = j = p$:

Let Z_a be the number of forms for a branch $a^\smile$. Let k be a positive nonzero integer. Thus

$$\vec{k!} = \sum_{m=0}^{k-1} (k - m) \quad (10)$$

and

$$\overleftarrow{k!} = \sum_{m=1}^{k} (m) \quad (11)$$

In the column $i = j = p$ the sum forms are accepted term by term due to the sum $(Z_p)\vec{!}$ and discarded term by term due to the sum $(Z_p - 1)\overleftarrow{!}$, in the following way:

Z_p $accepted$, 1 $discarded$,

$Z_p - 1$ $accepted$,

2 $discarded$,

$Z_p - 2$ $accepted$, $\quad 3$ $discarded$,

$Z_p - 3$ $accepted$, $\dots$,
$Z_p - 3$ $discarded$,
3 $accepted$,

$Z_p - 2$ $discarded$, 2 $accepted$,

$Z_p - 1$ $discarded$, 1 $accepted$.

$$\{1\}_{rie-}^{\Sigma o} \equiv 1^o$$

$$\{2\}_{rie-}^{\Sigma o} \equiv 1^o + 1^-$$

$$\{3\}_{rie-}^{\Sigma o} \equiv 1^o + 1^- \cdot 2$$

$$\{4\}_{rie-}^{\Sigma o} \equiv 1^o + 1^- \cdot 3 \doteq 1^o + 2^- + 1^-$$

$$\{5\}_{rie-}^{\Sigma o} \equiv 1^o + 1^- \cdot 4 \doteq 1^o + 2^- + 1^- \cdot 2$$
$$\doteq 1^o + 2^- \cdot 2$$

Notice that

$$k\vec{!} + (k-1)\overset{\leftharpoonup}{!} = k^2 \quad (12)$$

and

$$k\vec{!} = \frac{k(k+1)}{2} \quad (13)$$

1) $v = 8$

2) $8 = 1 + 7 = 1 + 7 = 1 + 7 = 1 + 7$

3) , 4) and 5)

$$8 = 1 + 1 \cdot 7 = 1 + 2 + 1 \cdot 5$$
$$= 1 + 2 \cdot 2 + 1 \cdot 3$$
$$= 1 + 2 \cdot 3 + 1$$
$$= 1 + (1 + 1 \cdot 2) + 1 \cdot 4$$
$$= 1 + (1 + 2) + 1 \cdot 4$$
$$= 1 + (1 + 1 \cdot 2) + 2 + 1 \cdot 2$$
$$= 1 + (1 + 2) + 2 + 1 \cdot 2$$
$$= 1 + (1 + 1 \cdot 2) + 2 \cdot 2$$
$$= 1 + (1 + 2) + 2 \cdot 2$$
$$= 1 + (1 + 1 \cdot 2) \cdot 2 + 1$$
$$= 1 + (1 + 2) + (1 + 1 \cdot 2) + 1$$
$$= 1 + (1 + 2) \cdot 2 + 1$$

$$= 1 + (1 + 1 \cdot 3) + 1 \cdot 3$$

$$= 1 + (1 + 2 + 1) + 1 \cdot 3$$

$$= 1 + \left(1 + (1 + 1 \cdot 2)\right) + 1 \cdot 3$$

$$= 1 + \left(1 + (1 + 2)\right) + 1 \cdot 3$$

$$= 1 + (1 + 2 + 1) + 2 + 1$$

$$= 1 + \left(1 + (1 + 1 \cdot 2)\right) + 2 + 1$$

$$= 1 + \left(1 + (1 + 2)\right) + 2 + 1$$

$$= 1 + \left(1 + (1 + 1 \cdot 2)\right) + (1 + 1 \cdot 2)$$

$$= 1 + \left(1 + (1 + 2)\right) + (1 + 1 \cdot 2)$$

$$= 1 + \left(1 + (1 + 2)\right) + (1 + 2) .$$

6) $\{8\}_{rie-}^{\Sigma o} \equiv 1^o + 1^- \cdot 7 \doteq 1^o + 2^- + 1^- \cdot 5$

$$\doteq 1^o + 2^- \cdot 2 + 1^- \cdot 3$$

$$\doteq 1^o + 2^- \cdot 3 + 1^-$$

$$\doteq 1^o + 3^x + 1^- \cdot 4$$

$$\doteq 1^o + 3^- + 1^- \cdot 4$$

$$\doteq 1^o + 3^x + 2^- + 1^- \cdot 2$$

$$\doteq 1^o + 3^- + 2^- + 1^- \cdot 2$$

$$\doteq 1^o + 3^x + 2^- \cdot 2$$

$$\doteq 1^o + 3^- + 2^- \cdot 2$$

$$\doteq 1^o + 3^x \cdot 2 + 1^-$$

$$\doteq 1^o + 3^- + 3^x + 1^-$$

$$\doteq 1^o + 3^- \cdot 2 + 1^-$$

$$\doteq 1^o + 4^x + 1^- \cdot 3$$

$$\doteq 1^o + (1^* + 2^- + 1^-) + 1^- \cdot 3$$

$$\doteq 1^o + (1^* + 3^x) + 1^- \cdot 3$$

$$\doteq 1^o + 4^- + 1^- \cdot 3$$

$$\doteq 1^o + (1^* + 2^- + 1^-) + 2^- + 1^-$$

$$\doteq 1^o + (1^* + 3^x) + 2^- + 1^-$$

$$\doteq 1^o + 4^- + 2^- + 1^-$$

$$\doteq 1^o + (1^* + 3^x) + 3^x$$

$$\doteq 1^o + 4^- + 3^x$$

$$\doteq 1^o + 4^- + 3^-.$$

v. Series-reduced free trees example. The tree generation algorithm works the same as for the rooted trees case, but with additional conditions from both the series-reduced rooted trees case and free trees case.

1) $v = 10$

2) $10 = 1 + 9 = 1 + 9 = 1 + 9 = 1 + 9$

3) , 4) and 5)

$$10 = 1 + 1 \cdot 9 = 1 + 3 + 1 \cdot 6$$
$$= 1 + 3 \cdot 2 + 1 \cdot 3$$
$$= 1 + 3 \cdot 3$$

$$= 1 + 4 + 1 \cdot 5$$
$$= 1 + 4 + 3 + 1 \cdot 2$$
$$= 1 + 4 \cdot 2 + 1$$

$$= 1 + (1 + 1 \cdot 4) + 1 \cdot 4$$
$$= 1 + (1 + 3 + 1) + 1 \cdot 4$$
$$= 1 + (1 + 3 + 1) + 3 + 1$$

6)$\{10\}_{rie-}^{\Sigma o} \equiv 1^o + 1^- \cdot 9$

$$\doteq 1^o + 3^x + 1^- \cdot 6$$

$$\doteq 1^o + 3^x \cdot 2 + 1^- \cdot 3$$

$$\doteq 1^o + 3^x \cdot 3$$

$$\doteq 1^o + 4^x + 1^- \cdot 5$$

$$\doteq 1^o + 4^x + 3^x + 1^- \cdot 2$$

$$\doteq 1^o + 4^x \cdot 2 + 1^-$$

$$\doteq 1^o + 5^x + 1^- \cdot 4$$

$$\doteq 1^o + (1^* + 3^x + 1^-) + 1^- \cdot 4$$

$$\doteq 1^o + (1^* + 3^x + 1^-) + 3^x + 1^-$$

1.2.3.3 Information about the tree generation algorithm

The tree generation algorithm is based on partitions: in how many ways can the number $v - 1$ be expressed as sums of positive nonzero integers, when v is the size of a tree. Sum forms, which can be expressed by the plain language of partitions, correspond to ordinary partitions of the number $v - 1$, with the except for the root; The sum forms with brackets are an additional feature of trees, and are not specifically ordinary partitions of the number $v - 1$. However, the brackets do not in an arithmetical way change these partitions, so the notion that trees are partitions is still quite right.

What comes to depth levels, the root corresponds to first and internal vertices to second, third and so on. The depth level starters ensure that every combination is considered, without the possibility of accidentally changing the places of branches during the generation process. An example of depth levels (the number in the exponent of the root / internal vertex defines the depth level):

$$(17)^o_{rie+} = 1^{o1} + (1^{*2} + 3^x \cdot 2 + (1^{*3} + 1^- \cdot 4) + 2^-) + (1^{*2} + 1^-)$$

Notice that the depth levels are the same for two internal vertices when these are inside the same number of bracket pairs.

The generated forest is divided into columns. The column C_n defines the size of the largest branch $n^\sim$ in the sum form. This way for example the column C_2 never exists for the series-reduced rooted and series-reduced free trees cases. The dividing of the columns according to the largest branch in the sum form eases the construction of forests. In these columns the partitions are put into lexicographical order (from up to down), thus creating a working generation process.

1.2.3.4 Branch form generator and tree enumeration algorithm

Branch form generator is a recursive algorithm, which generates all the forms of the chosen branch $n^{\sim}$, thus enumerating the number Z_n, to be used by the tree enumeration algorithm. The tree enumeration algorithm, on the other hand, works the same as tree generation algorithm, but with some exceptions.

1.2.3.5 Branch form generator

1) Let there be a branch $n^{\sim}$.

2) Think of the forms of the branch being in the same column. The generation order is from up to down and a new form starts a new row. Act in the way defined by the tree generation algorithm and use lexicographic order.

3) Use the notation $m^{a.-b.}$ as much as possible.

4) Use the notation $2^-(a,b)$(rooted and free trees) and the notation $3^x(a,b)$(series-reduced rooted trees and series-reduced free trees) as much as possible.

1.2.3.6Tree enumeration algorithm

1) The tree enumeration algorithm works the same as the tree generation algorithm with the exception that the notations $m^{a.-b.}$, $2^-(a,b)$ and $3^x(a,b)$ are used as much as possible.

2) Combination coefficients (Z) come in brackets after every enumeration row.

1.2.3.7 Examples

i.   Rooted trees, v = 1 − 10. For this tree type $Z_v = r_v$, so one algorithm is enough.

$1^{1.} = 1^-$

$2^{1.} = 2^-$

$3^{1.-2.} = (1^* + 2^-(0,1) + 1^-(2,0))$

$4^{1.-2.} = (1^* + 2^-(0,1) + 1^-(3,1))$

$4^{3.-4.} = (1^* + 3^{1.-2.})$

$5^{1.-3.} = (1^* + 2^-(0,2) + 1^-(4,0))$

$5^{4.-5.} = (1^* + 3^{1.-2.} + 1^-)$

$5^{6.-9.} = (1^* + 4^{1.-4.})$

$6^{1.-3.} = (1^* + 2^-(0,2) + 1^-(5,1))$

$6^{4.-7.} = (1^* + 3^{1.-2.} + 2^-(0,1) + 1^-(2,0))$

$6^{8.-11.} = (1^* + 4^{1.-4.} + 1^-)$

$6^{12.-20.} = (1^* + 5^{1.-9.})$

$7^{1.-4.} = (1^* + 2^-(0,3) + 1^-(6,0))$

$$7^{5.-8.} = (1^* + 3^{1.-2.} + 2^-(0,1) + 1^-(3,1))$$

$$7^{9.-11.} = (1^* + 3^{1.-2.} \cdot 2)$$

$$7^{12.-19.} = (1^* + 4^{1.-4.} + 2^-(0,1) + 1^-(2,0))$$

$$7^{20.-28.} = (1^* + 5^{1.-9.} + 1^-)$$
$$7^{29.-48.} = (1^* + 6^{1.-20.})$$
$$8^{1.-4.} = (1^* + 2^-(0,3) + 1^-(7,1))$$
$$8^{5.-10.} = (1^* + 3^{1.-2.} + 2^-(0,2) + 1^-(4,0))$$
$$8^{11.-13.} = (1^* + 3^{1.-2.} \cdot 2 + 1^-)$$
$$8^{14.-21.} = (1^* + 4^{1.-4.} + 2^-(0,1) + 1^-(3,1))$$
$$8^{22.-29.} = (1^* + 4^{1.-4.} + 3^{1.-2.})$$
$$8^{30.-47.} = (1^* + 5^{1.-9.} + 2^-(0,1) + 1^-(2,0))$$
$$8^{48.-67.} = (1^* + 6^{1.-20.} + 1^-)$$
$$8^{68.-115.} = (1^* + 7^{1.-48.})$$
$$9^{1.-5.} = (1^* + 2^-(0,4) + 1^-(8,0))$$
$$9^{6.-11.} = (1^* + 3^{1.-2.} + 2^-(0,2) + 1^-(5,1))$$
$$9^{12.-17.} = (1^* + 3^{1.-2.} \cdot 2 + 2^-(0,1) +$$
$$1^-(2,0))$$
$$9^{18.-29.} = (1^* + 4^{1.-4.} + 2^-(0,2) + 1^-(4,0))$$
$$9^{30.-37.} = (1^* + 4^{1.-4.} + 3^{1.-2.} + 1^-)$$
$$9^{38.-47.} = (1^* + 4^{1.-4.} \cdot 2)$$
$$9^{48.-65.} = (1^* + 5^{1.-9.} + 2^-(0,1) + 1^-(3,1))$$
$$9^{66.-83.} = (1^* + 5^{1.-9.} + 3^{1.-2.})$$

$$9^{84.-123.} = (1^* + 6^{1.-20.} + 2^-(0,1) + 1^-(2,0))$$

$$9^{124.-171.} = (1^* + 7^{1.-48.} + 1^-)$$

$$9^{172.-286.} = (1^* + 8^{1.-115.})$$

$$10^{1.-5.} = (1^* + 2^-(0,4) + 1^-(9,1))$$

$$10^{6.-13.} = (1^* + 3^{1.-2.} + 2^-(0,3) + 1^-(6,0))$$

$$10^{14.-19.} = (1^* + 3^{1.-2.} \cdot 2 + 2^-(0,1) + 1^-(3,1))$$

$$10^{20.-23.} = (1^* + 3^{1.-2.} \cdot 3)$$

$$10^{24.-35.} = (1^* + 4^{1.-4.} + 2^-(0,2) + 1^-(5,1))$$

$$10^{36.-51.} = (1^* + 4^{1.-4.} + 3^{1.-2.} + 2^-(0,1) + 1^-(2,0))$$

$$10^{52.-61.} = (1^* + 4^{1.-4.} \cdot 2 + 1^-)$$

$$10^{62.-88.} = (1^* + 5^{1.-9.} + 2^-(0,2) + 1^-(4,0))$$

$$10^{89.-106.} = (1^* + 5^{1.-9.} + 3^{1.-2.} + 1^-)$$

$$10^{107.-142.} = (1^* + 5^{1.-9.} + 4^{1.-4.})$$

$$10^{143.-182.} = (1^* + 6^{1.-20.} + 2^-(0,1) + 1^-(3,1))$$

$$10^{183.-222.} = (1^* + 6^{1.-20.} + 3^{1.-2.})$$

$$10^{223.-318.} = (1^* + 7^{1.-48.} + 2^-(0,1) + 1^-(2,0))$$

$$10^{319.-433.} = (1^* + 8^{1.-115.} + 1^-)$$

$$10^{434.-719.} = (1^* + 9^{1.-286.})$$

Thus

v:	1	2	3	4	5
r_v:	1	1	2	4	9
Number of algorithm rows:	1	1	1	2	3

v:	6	7	8	9	10
r_v:	20	48	115	286	719
Number of algorithm rows:	4	6	8	11	15

ii. Series-reduced rooted trees, v = 1 –
10. Now also the tree enumeration
algorithm is needed.

$1^{1.} = 1^-$

$3^{1.} = 3^x$

$4^{1.} = 4^x$

$5^{1.-2.} = (1^* + 3^x(0,1) + 1^-(4,1))$

$6^{1.-2.} = (1^* + 3^x(0,1) + 1^-(5,2))$

$6^{3.} = (1^* + 4^x + 1^-)$

$7^{1.-3.} = (1^* + 3^x(0,2) + 1^-(6,0))$

$7^{4.} = (1^* + 4^x + 1^- \cdot 2)$

$7^{5.-6.} = (1^* + 5^{1.-2.} + 1^-)$

$8^{1.-3.} = (1^* + 3^x(0,2) + 1^-(7,1))$

$8^{4.-5.} = (1^* + 4^x + 3^x(0,1) + 1^-(3,0))$

$8^{6.-7.} = (1^* + 5^{1.-2.} + 1^- \cdot 2)$

$$8^{8.-10.} = (1^* + 6^{1.-3.} + 1^-)$$

$$9^{1.-3.} = (1^* + 3^x(0,2) + 1^-(8,2))$$

$$9^{4.-5.} = (1^* + 4^x + 3^x(0,1) + 1^-(4,1))$$

$$9^{6.} = (1^* + 4^x \cdot 2)$$

$$9^{7.-10.} = (1^* + 5^{1.-2.} + 3^x(0,1) + 1^-(3,0))$$

$$9^{11.-13.} = (1^* + 6^{1.-3.} + 1^- \cdot 2)$$

$$9^{14.-19.} = (1^* + 7^{1.-6.} + 1^-)$$

$$\{1\}_{rie+}^{\Sigma o} \equiv 1^o(Z = 1)$$

$$\{2\}_{rie+}^{\Sigma o} \equiv 1^o + 1^-(Z = 1)$$

$$\{3\}_{rie+}^{\Sigma o} \equiv [1^o + 1^- \cdot 2](Z = 0) \text{ (sum form}$$
discarded)

$$\{4\}_{rie+}^{\Sigma o} \equiv 1^o + 3^x(0,1) + 1^-(3,0)(Z = 2)$$

$${\{5\}}_{rie+}^{\Sigma o} \equiv 1^o + 1^- \cdot 4(Z = 1) \doteq 1^o + 4^x(Z = 1)$$

$${\{6\}}_{rie+}^{\Sigma o} \equiv 1^o + 3^x(0,1) + 1^-(5,2)(Z = 2)$$

$$\doteq 1^o + 5^{1.-2.}(Z = 2)$$

$${\{7\}}_{rie+}^{\Sigma o} \equiv 1^o + 3^x(0,1) + 1^-(6,3)(Z = 2)$$

$$\doteq 1^o + 4^x + 1^- \cdot 2(Z = 1)$$

$$\doteq 1^o + 6^{1.-3.}(Z = 3)$$

$${\{8\}}_{rie+}^{\Sigma o} \equiv 1^o + 3^x(0,2) + 1^-(7,1)(Z = 3)$$

$$\doteq 1^o + 4^x + 1^- \cdot 3(Z = 1)$$

$$\doteq 1^o + 5^{1.-2.} + 1^- \cdot 2(Z = 2)$$

$$\doteq 1^o + 7^{1.-6.}(Z = 6)$$

$$\{9\}_{rie+}^{\Sigma o} \equiv 1^o + 3^x(0,2) + 1^-(8,2)(Z = 3)$$

$$\doteq 1^o + 4^x + 3^x(0,1) + 1^-(4,1)(Z = 2)$$

$$\doteq 1^o + 5^{1.-2.} + 1^- \cdot 3(Z = 2)$$

$$\doteq 1^o + 6^{1.-3.} + 1^- \cdot 2(Z = 3)$$

$$\doteq 1^o + 8^{1.-10.}(Z = 10)$$

$$\{10\}_{rie+}^{\Sigma o} \equiv 1^o + 3^x(0,3) + 1^-(9,0)(Z = 4)$$

$$\doteq 1^o + 4^x + 3^x(0,1) + 1^-(5,2)(Z = 2)$$

$$\doteq 1^o + 4^x \cdot 2 + 1^-(Z = 1)$$

$$\doteq 1^o + 5^{1.-2.} + 3^x(0,1) + 1^-(4,1)(Z = 4)$$

$$\doteq 1^o + 6^{1.-3.} + 1^- \cdot 3(Z = 3)$$

$$\doteq 1^o + 7^{1.-6.} + 1^- \cdot 2(Z = 6)$$

$$\doteq 1^o + 9^{1.-19.}(Z = 19)$$

Thus

$$s_1 = 1$$

$$s_2 = 1$$

$$s_3 = 0$$

$$s_4 = 2$$

$$s_5 = 1 + 1 = 2$$

$$s_6 = 2 + 2 = 4$$

$$s_7 = 2 + 1 + 3 = 6$$

$$s_8 = 3 + 1 + 2 + 6 = 12$$

$$s_9 = 3 + 2 + 2 + 3 + 10 = 20$$

$$s_{10} = 4 + 2 + 1 + 4 + 3 + 6 + 19 = 39$$

iii.   Free trees, v = 1 − 10. The generated branch forms in the rooted trees case can now be used. The sum forms in the middle column are not compactified for the sake of clarity.

$$\{1\}_{rie-}^{\Sigma o} \equiv 1^o (Z = 1)$$

$$\{2\}_{rie-}^{\Sigma o} \equiv 1^o + 1^-(Z = 1)$$

$$\{3\}_{rie-}^{\Sigma o} \equiv 1^o + 1^- \cdot 2(Z = 1)$$

$$\{4\}_{rie-}^{\Sigma o} \equiv 1^o + 2^-(0,1) + 1^-(3,1)(Z = 2)$$

$$\{5\}_{rie-}^{\Sigma o} \equiv 1^o + 2^-(0,2) + 1^-(4,0)(Z = 3)$$

$$\{6\}_{rie-}^{\Sigma o} \equiv 1^o + 2^-(0,2) + 1^-(5,1)(Z = 3)$$

$$\doteq 1^o + 3^{1.-2.} + 1^- \cdot 2(Z = 2)$$

$$\doteq 1^o + 3^{2.} + 2^-(Z = 1)$$

$$\{7\}_{rie-}^{\Sigma o} \equiv 1^o + 2^-(0,3) + 1^-(6,0)(Z = 4)$$

$$\doteq 1^o + 3^{1.-2.} + 2^-(0,1) + 1^-(3,1)(Z = 4)$$

$$\doteq 1^o + 3^{1.-2.} \cdot 2(Z = 3)$$

$$\{8\}_{rie-}^{\Sigma o} \equiv 1^o + 2^-(0,3) + 1^-(7,1)(Z = 4)$$

$$\doteq 1^o + 3^{1.-2.} + 2^-(0,2) + 1^-(4,0)(Z = 6)$$

$$\doteq 1^o + 3^{1.-2.} \cdot 2 + 1^-(Z = 3)$$

$$\doteq 1^o + 4^{1.-4.} + 1^- \cdot 3(Z = 4)$$

$$\doteq 1^o + 4^{2.-4.} + 2^- + 1^-(Z = 3)$$

$$\doteq 1^o + 4^{3.-4.} + 3^{1.}(Z = 2)$$

$$\doteq 1^o + 4^{4.} + 3^{2.}(Z = 1)$$

$$\{9\}_{rie-}^{\Sigma o} \equiv 1^o + 2^-(0,4) + 1^-(8,0)(Z = 5)$$

$$\dot{=} 1^o + 3^{1.-2.} + 2^-(0,2) +$$
$$1^-(5,1)(Z = 6)$$

$$\dot{=} 1^o + 3^{1.-2.} \cdot 2 + 2^-(0,1) +$$
$$1^-(2,0)(Z = 6)$$

$$\dot{=} 1^o + 4^{1.-4.} + 2^-(0,2) +$$
$$1^-(4,0)(Z = 12)$$

$$\dot{=} 1^o + 4^{1.-4.} + 3^{1.-2.} + 1^-(Z = 8)$$

$$\dot{=} 1^o + 4^{1.-4.} \cdot 2(Z = 10)$$

$$\{10\}_{rie-}^{\Sigma o} \equiv 1^o + 2^-(0,4) + 1^-(9,1)(Z = 5)$$

$$\dot{=} 1^o + 3^{1.-2.} + 2^-(0,3) +$$
$$1^-(6,0)(Z = 8)$$

$$\dot{=} 1^o + 3^{1.-2.} \cdot 2 + 2^-(0,1) +$$
$$1^-(3,1)(Z = 6)$$

$$\dot{=} 1^o + 3^{1.-2.} \cdot 3(Z = 4)$$

$$\doteq 1^o + 4^{1.-4.} + 2^-(0,2) + 1^-(5,1)(Z = 12)$$

$$\doteq 1^o + 4^{1.-4.} + 3^{1.-2.} + 2^-(0,1) + 1^-(2,0)(Z = 16)$$

$$\doteq 1^o + 4^{1.-4.} \cdot 2 + 1^-(Z = 10)$$

$$\doteq 1^o + 5^{1.-9.} + 1^- \cdot 4(Z = 9)$$

$$\doteq 1^o + 5^{2.-9.} + 2^- + 1^- \cdot 2(Z = 8)$$

$$\doteq 1^o + 5^{3.-9.} + 2^- \cdot 2(Z = 7)$$

$$\doteq 1^o + 5^{4.-9.} + 3^{1.} + 1^-(Z = 6)$$

$$\doteq 1^o + 5^{5.-9.} + 3^{2.} + 1^-(Z = 5)$$

$$\doteq 1^o + 5^{6.-9.} + 4^{1.}(Z = 4)$$

$$\doteq 1^o + 5^{7.-9.} + 4^{2.}(Z = 3)$$

$$\doteq 1^o + 5^{8.-9.} + 4^{3.}(Z = 2)$$

$$\doteq 1^o + 5^{9.} + 4^{4.}(Z = 1)$$

Thus

$$t_1 = 1$$

$$t_2 = 1$$

$$t_3 = 1$$

$$t_4 = 2$$

$$t_5 = 3$$

$$t_6 = 3 + 2 + 1 = 6$$

$$t_7 = 4 + 4 + 3 = 11$$

$$t_8 = 4 + 6 + 3 + 4 + 3 + 2 + 1 = 23$$

$$t_9 = 5 + 6 + 6 + 12 + 8 + 10 = 47$$

$$t_{10} = 5 + 8 + 6 + 4 + 12 + 16 + 10 + 9 + 8 + 7 + 6 + 5 + 4 + 3 + 2 + 1 = 106$$

iv.   Series-reduced free trees, v = 1 –
10. The generated branch forms in
the series-reduced rooted trees
case can now be used. The sum
forms in the middle column are not
compactified for the sake of clarity.

$\{1\}^{\Sigma o}_{rie-} \equiv 1^o (Z = 1)$

$\{2\}^{\Sigma o}_{rie-} \equiv 1^o + 1^- (Z = 1)$

$\{3\}^{\Sigma o}_{rie-} \equiv [1^o + 1^- \cdot 2](Z = 0)$(Sum
form discarded)

$\{4\}^{\Sigma o}_{rie-} \equiv 1^o + 1^- \cdot 3(Z = 1)$

$\{5\}^{\Sigma o}_{rie-} \equiv 1^o + 1^- \cdot 4(Z = 1)$

$\{6\}^{\Sigma o}_{rie-} \equiv 1^o + 3^x (0,1) + 1^- (5,2)(Z = 2)$

$$\{7\}_{rie-}^{\Sigma o} \equiv 1^o + 3^x(0,1) + 1^-(6,3)(Z = 2)$$

$$\{8\}_{rie-}^{\Sigma o} \equiv 1^o + 3^x(0,2) + 1^-(7,1)(Z = 3)$$

$$\doteq 1^o + 4^x + 1^- \cdot 3(Z = 1)$$

$$\{9\}_{rie-}^{\Sigma o} \equiv 1^o + 3^x(0,2) + 1^-(8,2)(Z = 3)$$

$$\doteq 1^o + 4^x + 3^x(0,1) + 1^-(4,1)(Z = 2)$$

$$\{10\}_{rie-}^{\Sigma o} \equiv 1^o + 3^x(0,3) + 1^-(9,0)(Z = 4)$$

$$\doteq 1^o + 4^x + 3^x(0,1) + 1^-(5,2)(Z = 2)$$

$$\doteq 1^o + 4^x \cdot 2 + 1^-(Z = 1)$$

$$\doteq 1^o + 5^{1.-2.} + 1^- \cdot 4(Z = 2)$$

$$\doteq 1^o + 5^{2.} + 3^x + 1^-(Z = 1)$$

Thus

$$h_1 = 1$$

$$h_2 = 1$$

$$h_3 = 0$$

$$h_4 = 1$$

$$h_5 = 1$$

$$h_6 = 2$$

$$h_7 = 2$$

$$h_8 = 3 + 1 = 4$$

$$h_9 = 3 + 2 = 5$$

$$h_{10} = 4 + 2 + 1 + 2 + 1 = 10$$

Let us show now a more intensive enumeration example for series-reduced free trees.

$$1^{1.} = 1^-$$

$$3^{1.} = 3^x$$

$$4^{1.} = 4^x$$

$$5^{1.-2.} = (1^* + 3^x(0,1) + 1^-(4,1))$$

$$6^{1.-2.} = (1^* + 3^x(0,1) + 1^-(5,2))$$

$$6^{3.} = (1^* + 4^x + 1^-)$$

$$7^{1.-3.} = (1^* + 3^x(0,2) + 1^-(6,0))$$

$$7^{4.} = (1^* + 4^x + 1^- \cdot 2)$$

$$7^{5.-6.} = (1^* + 5^{1.-2.} + 1^-)$$

$$8^{1.-3.} = (1^* + 3^x(0,2) + 1^-(7,1))$$

$$8^{4.-5.} = (1^* + 4^x + 3^x(0,1) + 1^-(3,0))$$

$$8^{6.-7.} = (1^* + 5^{1.-2.} + 1^- \cdot 2)$$

$$8^{8.-10.} = (1^* + 6^{1.-3.} + 1^-)$$

$$9^{1.-3.} = (1^* + 3^x(0,2) + 1^-(8,2))$$

$$9^{4.-5.} = (1^* + 4^x + 3^x(0,1) + 1^-(4,1))$$

$$9^{6.} = (1^* + 4^x \cdot 2)$$

$$9^{7.-10.} = (1^* + 5^{1.-2.} + 3^x(0,1) + 1^-(3,0))$$

$$9^{11.-13.} = (1^* + 6^{1.-3.} + 1^- \cdot 2)$$

$$9^{14.-19.} = (1^* + 7^{1.-6.} + 1^-)$$

$$10^{1.-4.} = (1^* + 3^x(0,3) + 1^-(9,0))$$

$$10^{5.-6.} = (1^* + 4^x + 3^x(0,1) + 1^-(5,2))$$

$$10^{7.} = (1^* + 4^x \cdot 2 + 1^-)$$

$$10^{8.-11.} = (1^* + 5^{1.-2.} + 3^x(0,1) + 1^-(4,1))$$

$$10^{12.-13.} = (1^* + 5^{1.-2.} + 4^x)$$

$$10^{14.-19.} = (1^* + 6^{1.-3.} + 3^x(0,1) + 1^-(3,0))$$

$$10^{20.-25.} = (1^* + 7^{1.-6.} + 1^- \cdot 2)$$

$$10^{26.-35.} = (1^* + 8^{1.-10.} + 1^-)$$

v = 20.

As in the column $C_{20/2} = C_{10}$ there are $(Z_{10})\vec{!} = 35\vec{!} =$
$$\frac{35 \cdot (1+35)}{2} = \frac{35 \cdot 36}{2} = \frac{(30+5)(30+6)}{2} = \frac{900+180+150+30}{2} =$$
$\frac{1260}{2} = 630$ sum forms, this column can be left out from the next enumeration process, while the number 630 is added to the result.

$$\{20\}_{rie-}^{\Sigma o} \equiv 1^o + 3^x(0,6) + 1^-(19,1)(Z = 7)$$

$$\doteq 1^o + 4^x + 3^x(0,5) + 1^-(15,0)(Z = 6)$$

$$\doteq 1^o + 4^x \cdot 2 + 3^x(0,3) + 1^-(11,2)(Z = 4)$$

$$\doteq 1^o + 4^x \cdot 3 + 3^x(0,2) + 1^-(7,1)(Z = 3)$$

$$\doteq 1^o + 4^x \cdot 4 + 3^x(0,1) + 1^-(3,0)(Z = 2)$$

$$\doteq 1^o + 5^{1.-2.} + 3^x(0,4) + 1^-(14,2)(Z = 10)$$

$$\doteq 1^o + 5^{1.-2.} + 4^x + 3^x(0,3) + 1^-(10,1)(Z = 8)$$

$$\doteq 1^o + 5^{1.-2.} + 4^x \cdot 2 + 3^x(0,2) + 1^-(6,0)(Z = 6)$$

$$\doteq 1^o + 5^{1.-2.} + 4^x \cdot 3 + 1^- \cdot 2(Z = 2)$$

$$\doteqdot 1^o + 5^{1.-2.} \cdot 2 + 3^x(0,3) + 1^-(9,0)(Z = 12)$$

$$\doteqdot 1^o + 5^{1.-2.} \cdot 2 + 4^x + 3^x(0,1) + 1^-(5,2)(Z = 6)$$

$$\doteqdot 1^o + 5^{1.-2.} \cdot 2 + 4^x \cdot 2 + 1^-(Z = 3)$$

$$\doteqdot 1^o + 5^{1.-2.} \cdot 3 + 3^x(0,1) + 1^-(4,1)(Z = 8)$$

$$\doteqdot 1^o + 5^{1.-2.} \cdot 3 + 4^x(Z = 4)$$

$$\doteqdot 1^o + 6^{1.-3.} + 3^x(0,4) + 1^-(13,1)(Z = 15)$$

$$\doteqdot 1^o + 6^{1.-3.} + 4^x + 3^x(0,3) + 1^-(9,0)(Z = 12)$$

$$\doteqdot 1^o + 6^{1.-3.} + 4^x \cdot 2 + 3^x(0,1) + 1^-(5,2)(Z = 6)$$

$$\doteqdot 1^o + 6^{1.-3.} + 4^x \cdot 3 + 1^-(Z = 3)$$

$$\doteqdot 1^o + 6^{1.-3.} + 5^{1.-2.} + 3^x(0,2) + 1^-(8,2)(Z = 18)$$

$$\doteqdot 1^o + 6^{1.-3.} + 5^{1.-2.} + 4^x + 3^x(0,1) + 1^-(4,1)(Z = 12)$$

$$\doteqdot 1^o + 6^{1.-3.} + 5^{1.-2.} + 4^x \cdot 2(Z = 6)$$

$$\doteqdot 1^o + 6^{1.-3.} + 5^{1.-2.} \cdot 2 + 3^x(0,1) + 1^-(3,0)(Z = 18)$$

$$\doteqdot 1^o + 6^{1.-3.} \cdot 2 + 3^x(0,2) + 1^-(7,1)(Z = 18)$$

$$\doteq 1^o + 6^{1.-3.} \cdot 2 + 4^x + 3^x(0,1) + 1^-(3,0)(Z = 12)$$

$$\doteq 1^o + 6^{1.-3.} \cdot 2 + 5^{1.-2.} + 1^- \cdot 2(Z = 12)$$

$$\doteq 1^o + 6^{1.-3.} \cdot 3 + 1^-(Z = 10)$$

$$\doteq 1^o + 7^{1.-6.} + 3^x(0,4) + 1^-(12,0)(Z = 30)$$

$$\doteq 1^o + 7^{1.-6.} + 4^x + 3^x(0,2) + 1^-(8,2)(Z = 18)$$

$$\doteq 1^o + 7^{1.-6.} + 4^x \cdot 2 + 3^x(0,1) + 1^-(4,1)(Z = 12)$$

$$\doteq 1^o + 7^{1.-6.} + 4^x \cdot 3(Z = 6)$$

$$\doteq 1^o + 7^{1.-6.} + 5^{1.-2.} + 3^x(0,2) + 1^-(7,1)(Z = 36)$$

$$\doteq 1^o + 7^{1.-6.} + 5^{1.-2.} + 4^x + 3^x(0,1) + 1^-(3,0)(Z = 24)$$

$$\doteq 1^o + 7^{1.-6.} + 5^{1.-2.} \cdot 2 + 1^- \cdot 2(Z = 18)$$

$$\doteq 1^o + 7^{1.-6.} + 6^{1.-3.} + 3^x(0,2) + 1^-(6,0)(Z = 54)$$

$$\doteq 1^o + 7^{1.-6.} + 6^{1.-3.} + 4^x + 1^- \cdot 2(Z = 18)$$

$$\doteq 1^o + 7^{1.-6.} + 6^{1.-3.} + 5^{1.-2.} + 1^-(Z = 36)$$

$$\doteq 1^o + 7^{1.-6.} + 6^{1.-3.} \cdot 2(Z = 36)$$

$$\doteq 1^o + 7^{1.-6.} \cdot 2 + 3^x(0,1) + 1^-(5,2)(Z = 42)$$

$$\doteq 1^o + 7^{1.-6.} \cdot 2 + 4^x + 1^-(Z = 21)$$

$$\doteq 1^o + 7^{1.-6.} \cdot 2 + 5^{1.-2.}(Z = 42)$$

$$\doteq 1^o + 8^{1.-10.} + 3^x(0,3) + 1^-(11,2)(Z = 40)$$

$$\doteq 1^o + 8^{1.-10.} + 4^x + 3^x(0,2) + 1^-(7,1)(Z = 30)$$

$$\doteq 1^o + 8^{1.-10.} + 4^x \cdot 2 + 3^x(0,1) + 1^-(3,0)(Z = 20)$$

$$\doteq 1^o + 8^{1.-10.} + 5^{1.-2.} + 3^x(0,2) + 1^-(6,0)(Z = 60)$$

$$\doteq 1^o + 8^{1.-10.} + 5^{1.-2.} + 4^x + 1^- \cdot 2(Z = 20)$$

$$\doteq 1^o + 8^{1.-10.} + 5^{1.-2.} \cdot 2 + 1^-(Z = 30)$$

$$\doteq 1^o + 8^{1.-10.} + 6^{1.-3.} + 3^x(0,1) + 1^-(5,2)(Z = 60)$$

$$\doteq 1^o + 8^{1.-10.} + 6^{1.-3.} + 4^x + 1^-(Z = 30)$$

$$\doteq 1^o + 8^{1.-10.} + 6^{1.-3.} + 5^{1.-2.}(Z = 60)$$

$$\doteq 1^o + 8^{1.-10.} + 7^{1.-6.} + 3^x(0,1) + 1^-(4,1)(Z = 120)$$

$$\doteq 1^o + 8^{1.-10.} + 7^{1.-6.} + 4^x(Z = 60)$$

$$\doteq 1^o + 8^{1.-10.} \cdot 2 + 3^x(0,1) + 1^-(3,0)(Z = 110)$$

$$\doteq 1^o + 9^{1.-19.} + 3^x(0,3) + 1^-(10,1)(Z = 76)$$

$$\doteq 1^o + 9^{1.-19.} + 4^x + 3^x(0,2) + 1^-(6,0)(Z = 57)$$

$$\doteq 1^o + 9^{1.-19.} + 4^x \cdot 2 + 1^- \cdot 2(Z = 19)$$

$$\doteq 1^o + 9^{1.-19.} + 5^{1.-2.} + 3^x(0,1) + 1^-(5,2)(Z = 76)$$

$$\doteq 1^o + 9^{1.-19.} + 5^{1.-2.} + 4^x + 1^-(Z = 38)$$

$$\doteq 1^o + 9^{1.-19.} + 5^{1.-2.} \cdot 2(Z = 57)$$

$$\doteq 1^o + 9^{1.-19.} + 6^{1.-3.} + 3^x(0,1) + 1^-(4,1)(Z = 114)$$

$$\doteq 1^o + 9^{1.-19.} + 6^{1.-3.} + 4^x(Z = 57)$$

$$\doteq 1^o + 9^{1.-19.} + 7^{1.-6.} + 3^x(0,1) + 1^-(3,0)(Z = 228)$$

$$\doteq 1^o + 9^{1.-19.} + 8^{1.-10.} + 1^- \cdot 2(Z = 190)$$

$$\doteq 1^o + 9^{1.-19.} \cdot 2 + 1^-(Z = 190)$$

Thus

$$h_{20} = 7 + 6 + 4 + 3 + 2 + 10 + 8 + 6 + 2 + 12 + 6 + 3 + 8 + 4 + 15 + 12 + 6 + 3 + 18 + 12 + 6 + 18 + 18 + 12 + 12 + 10 + 30 + 18 + 12 + 6 + 36 + 24 + 18 + 54 + 18 + 36 + 36 + 42 + 21 + 42 + 40 + 30 + 20 + 60 + 20 + 30 + 60 + 30 + 60 + 120 + 60 + 110 + 76 + 57 + 19 + 76 + 38 + 57 + 114 + 57 + 228 + 190 + 190 + 630 = 2988.$$

So

$$h_{20} = 2988 \, .$$

1.3 Enumeration formula for rooted trees

Due to the link between trees and partitions, the following formula can be constructed:

$$r_n = C(P(n-1)) \tag{14}$$

, where C is an entity that takes in a list P of all partitions of the integer n − 1 ($n \geq 2$, with $r_1 = 1$), transforms the partitions into products of binomial coefficients in the same way that the formula (8) does, and then sums up the products and outputs the sum as the quantity for rooted trees with tree size n.

Example:

$$P(6) \equiv 1 \cdot 6 = 2 + 1 \cdot 4 = 2 \cdot 2 + 1 \cdot 2 = 2 \cdot 3 = 3 + 1 \cdot 3 = 3 + 2 + 1 = 3 \cdot 2 = 4 + 1 \cdot 2 = 4 + 2 = 5 + 1 = 6$$

$$C(P(6)) = \binom{r_1 + 6 - 1}{6} + \binom{r_2 + 1 - 1}{1}\binom{r_1 + 4 - 1}{4}$$
$$+ \binom{r_2 + 2 - 1}{2}\binom{r_1 + 2 - 1}{2}$$
$$+ \binom{r_2 + 3 - 1}{3}$$
$$+ \binom{r_3 + 1 - 1}{1}\binom{r_1 + 3 - 1}{3}$$
$$+ \binom{r_3 + 1 - 1}{1}\binom{r_2 + 1 - 1}{1}\binom{r_1 + 1 - 1}{1}$$
$$+ \binom{r_3 + 2 - 1}{2}$$
$$+ \binom{r_4 + 1 - 1}{1}\binom{r_1 + 2 - 1}{2}$$
$$+ \binom{r_4 + 1 - 1}{1}\binom{r_2 + 1 - 1}{1}$$
$$+ \binom{r_5 + 1 - 1}{1}\binom{r_1 + 1 - 1}{1}$$
$$+ \binom{r_6 + 1 - 1}{1}$$

$$C(P(6)) = \binom{r_1 + 5}{6} + \binom{r_2}{1}\binom{r_1 + 3}{4}$$
$$+ \binom{r_2 + 1}{2}\binom{r_1 + 1}{2} + \binom{r_2 + 2}{3}$$
$$+ \binom{r_3}{1}\binom{r_1 + 2}{3} + \binom{r_3}{1}\binom{r_2}{1}\binom{r_1}{1}$$
$$+ \binom{r_3 + 1}{2} + \binom{r_4}{1}\binom{r_1 + 1}{2} + \binom{r_4}{1}\binom{r_2}{1}$$
$$+ \binom{r_5}{1}\binom{r_1}{1} + \binom{r_6}{1}$$

$$C(P(6)) = \binom{6}{6} + \binom{1}{1}\binom{4}{4} + \binom{2}{2}\binom{2}{2} + \binom{3}{3} + \binom{2}{1}\binom{3}{3}$$
$$+ \binom{2}{1}\binom{1}{1}\binom{1}{1} + \binom{3}{2} + \binom{4}{1}\binom{2}{2} + \binom{4}{1}\binom{1}{1}$$
$$+ \binom{9}{1}\binom{1}{1} + \binom{20}{1}$$

$$C(P(6)) = 1 + 1 \cdot 1 + 1 \cdot 1 + 1 + 2 \cdot 1 + 2 \cdot 1 \cdot 1 +$$
$$3 + 4 \cdot 1 + 4 \cdot 1 + 9 \cdot 1 + 20$$

$$C(P(6)) = 1 + 1 + 1 + 1 + 2 + 2 + 3 + 4 + 4 + 9 +$$
$$20 = 48$$

$$C(P(6)) = 48$$

$$r_7 = 48.$$

1.4 Treespeak

Treespeak is an idea of a world where treespeak users, also known as gifted users, use a verbal version of vertex edge algebra to create and destroy. This verbal version, combined with the order defined by the tree generation algorithm, forms treespeak, which gives a possibility to manipulate the graph field that surrounds the user. The manipulation especially includes the trees, and these together with graphs work as the framework for the structures of chemical molecules, atoms, and trajectories in physics. These trees and graphs also have a connection to data structures of program languages, so hacking / programming is possible, in addition to chemistry and physics. Mathematics is however the core of this all, and it can be used to improve and modify one's abilities in treespeak.

The ability of a gifted user to use graphs and trees is measured as a number called tree size metric, or tsm. The higher the tsm, the larger the trees one can use and thus the more complicated and possibly deadly the spells can be. A battle – or tree generation process – between two treespeak users usually culminates to one of the users as winning the battle, and usually this user has a higher tsm. According to estimates, the tsm of the most talented users is over one hundred, but according to the rules of math the tsm should not have any upper bound. The mind of a human may be limited, but what about an entity higher than human? Maybe one theorist knows...

You can check the mathematical tool / game Treespeak in my website www.matladpi.com. Or, alternatively, go to the website https://www.simmer.io/@matladpi/treespeak for the same tool / game with a description. Also, if you are interested, feel free to join the treespeak subreddit r/mathematicaltreespeak in the website www.reddit.com.

1.5 Axiomatic Peano-like system for rooted trees

A tree is a mathematical object in which there is a root / virtual root and possible branches attached to it. Let R be the sequence of all rooted trees and $r(i)$ be a rooted tree with a positive integer index i in the sequence. Let S be the sequence of all sum forms for rooted trees and $s(i)$ be a sum form with a positive integer index i in the sequence. Let also G be the sequence of all graphical forms for rooted trees and $g(i)$ be a graphical form with a positive integer index i in the sequence.

The theorem of sum forms establishes the use of sum forms for rooted trees such that: if $s(i)$ is the sum form and $g(i)$ is the graphical form of a rooted tree $r(i)$, then one can change between the graphical forms $g(i)$ and sum forms $s(i)$ freely. The tree generation algorithm establishes the use of the successor operator S_+^m , $m \in \mathbb{N} \setminus \{0\}$, with defining the order in which the rooted trees are in the sequences R, S and G.

1) $s(1) = 1$ and $1 \in S$

2) For all rooted trees and indexes $i \in \mathbb{N} \setminus \{0\}$, $r(i) = r(i)$, $s(i) = s(i)$ and $g(i) = g(i)$

3) For rooted tree pairs $(r(i), r(j))$ and corresponding sum form pairs $(s(i), s(j))$ and graphical form pairs $(g(i), g(j))$:
   - If $r(i) = r(j)$, then $r(j) = r(i)$ and $i = j$
   - If $s(i) = s(j)$, then $s(j) = s(i)$ and $i = j$
   - If $g(i) = g(j)$, then $g(j) = g(i)$ and $i = j$

4) For rooted tree triplets $(r(i), r(j), r(k))$ and corresponding sum form triplets $(s(i), s(j), s(k))$ and graphical form triplets $(g(i), g(j), g(k))$:
   - If $r(i) = r(j)$ and $r(j) = r(k)$, then $r(i) = r(k)$ and $i = j = k$
   - If $s(i) = s(j)$ and $s(j) = s(k)$, then $s(i) = s(k)$ and $i = j = k$
   - If $g(i) = g(j)$ and $g(j) = g(k)$, then $g(i) = g(k)$ and $i = j = k$

5) For rooted tree pairs $(r(i), r(j))$ and corresponding sum form pairs $(s(i), s(j))$ and graphical form pairs $(g(i), g(j))$:

- If $r(j) \in R$ and $r(i) = r(j)$, then also $r(i) \in R$
- If $s(j) \in S$ and $s(i) = s(j)$, then also $s(i) \in S$
- If $g(j) \in G$ and $g(i) = g(j)$, then also $g(i) \in G$

6) Let S_+^m be a successor operator. For every arbitrary rooted tree $r(i)$ and corresponding sum form $s(i)$ and graphical form $g(i)$:

- $S_+^1(r(i)) = S_+(r(i)) = r(i+1) \in R$
- $S_+^1(s(i)) = S_+(s(i)) = s(i+1) \in S$
- $S_+^1(g(i)) = S_+(g(i)) = g(i+1) \in G$

7) For rooted tree pairs $(r(i), r(j))$ and corresponding sum form pairs $(s(i), s(j))$ and graphical form pairs $(g(i), g(j))$:

- $r(i) = r(j)$ if and only if $S_+^m(r(i)) = r(i+m) = S_+^m(r(j)) = r(j+m)$

- $s(i) = s(j)$ if and only if $S_+^m(s(i)) = s(i + m) = S_+^m(s(j)) = s(j + m)$
- $g(i) = g(j)$ if and only if $S_+^m(g(i)) = g(i + m) = S_+^m(g(j)) = g(j + m)$

8) For all sum forms $s(i)$ and indexes $i \in \mathbb{N} \setminus \{0\}$, $S_+(s(i)) \neq 1$

9) If P is a set such that $s(1) = 1 \in P$ and for every sum form $s(i)$, $s(i)$ being in P implies that $S_+(s(i))$ is in P , then P contains every rooted tree.

10)   For every rooted tree pair $(r(i), r(j))$, if $|r(i)| > |r(j)|$, then $|S_+(r(i))| \geq |S_+(r(j))|$

11)   The order of the sequence S and thus the order of the sequences R and G are defined by the tree generation algorithm.

12) $s(i) + 0 = s(i)$

13) For every arbitrary rooted tree $r(i)$, if $|r(i)| = n$, then $r(i) \in R_n$, where R_n is a rooted tree forest, consisting of every rooted tree with tree size n; $|R_n| = r_n$. S_n and G_n are the corresponding sum form and graphical form of the said forest; If $|r(i)| = |s(i)| = |g(i)| = n$, then $s(i) \in S_n$ and $g(i) \in G_n$. $R_n(i)$ is the rooted tree with an index i in the forest of all the rooted trees with size n. Correspondingly, $S_n(i)$ is the sum form and $G_n(i)$ is the graphical form of that same tree.

14) The manipulation of a tree's spatial alignment does not change the properties of the tree. In other words, no new trees can be created with this procedure.

15) The manipulation of the graphical sizes of a tree's different parts does not change the properties of the tree. In other words, no new trees can be created with this procedure.

These axioms can be extended to symbolic vertex edge algebraic forms, also known as symbolic sum forms. As usual, the tree generation algorithm defines the order in the sequences R, S and G, and in the sequences R_n, S_n and G_n. For rooted trees,

$$s(1) = 1^o$$

$$s(i + 1) = S_+(s(i)) , i \geq 1$$

$$s(i + m) = S_+^m(s(i)) , i \geq 1, m \geq 1$$

$$S_+(1^o + n^-) = 1^o + 1^- \cdot (n + 1) , n \geq 0$$

$$S_+(1^o + 1^- \cdot n) = 1^o + 2^- + 1^- \cdot (n - 2) , n \geq 2$$

$$S_+(1^o + n^x) = 1^o + n^L , n \geq 3$$

$$S_+(1^o + n^L) = 1^o + n^{2L} , n \geq 5$$

$$S_+(1^o + n^Y) = 1^o + n^- , n \geq 3$$

$$S_+(1^o + n^\sim) = 1^o + S_+(n^\sim) , n \geq 3$$

, with the assumption that $n^\sim$ has at least two forms and the current one is not the last form.

$$S_+(n^{k\cdot}) = n^{(k+1).}$$

, with the assumption that $n \geq 3$ and $Z_n \geq k + 1$

$$S_+(n^{k_1\cdot} + n^{k_2\cdot}) = n^{(k_1+1).} + n^{k_2\cdot}$$

, with the assumption that $n \geq 3$, $Z_n \geq k_1 + 1$ and $k_1 \geq k_2$

$$S_+\left(n^{k_0\cdot} + \sum_{m=1}^{p}(n^{k_m\cdot})\right) = n^{(k_0+1).} + \sum_{m=1}^{p}(n^{k_m\cdot})$$

, with the assumption that $n \geq 3$, $Z_n \geq k_0 + 1$ and $k_0 \geq k_1 \geq k_2 \geq \cdots \geq k_{p-1} \geq k_p$

$$S_+\left(n_1^{a\cdot} + n_2^{b\cdot}\right) = n_1^{(a+1).} + n_2^{b\cdot}$$

, with the assumption that $n_1 \geq n_2 \geq 3$, $Z_{n_1} \geq a + 1$, $Z_{n_2} \geq b$.

1.6Possibly open problems

1. Create a general method which calculates an identity number for any sum form as an input, making it possible to calculate the general average identity number of a rooted tree / free tree with the aid of a computer.

Identity number is about the structure of a tree and is divided into two numbers:

1) The number $\ell_- \in [0, 1]$, $\ell_- \neq \dfrac{n-3}{n-1}$, $\dfrac{n-2}{n-1}$, $n \geq 3$, when n is the number of a tree's vertices and the numerator / denominator of ℓ_- is not canceled / expanded.

2) The number $\ell_x \in [0, 1]$, $\ell_x \neq \dfrac{1}{n-1}$, $\dfrac{2}{n-1}$, $n \geq 3$, when n is the number of a tree's vertices and the numerator / denominator of ℓ_- is not canceled / expanded.

From these, the rational number (finite tree) ℓ_- is about the longitudinal nature of the tree. The larger this number is, the more longitudinal the tree is.

The rational number (finite tree) ℓ_x, on the other hand, is about the tree's furcate nature. The larger this number is, the more furcate the tree is.

These numbers can also be turned into percentages, although the original value should be preserved. Namely, this number contains some original information about the tree, for example the number of its edges.

The number ℓ_- can be calculated in the following way:

Choose a vertex, which is connected to at least three other vertices (if there is no such vertex, then $\ell_- = 1$) with edges. The counting happens with going outwards from the vertex in question. Count the edges in all the longitudinal parts, in which there are at least two or more edges in a line. In these parts $d_i = 2 \ / \ d_r = 2$. When you have calculated the total number of edges in these parts, subtract the number of the longitudinal parts from the total number and then divide the result by the total number of the edges in the tree.

The number ℓ_x can be calculated in the following way:

Choose a vertex, which is connected to at least three other vertices (if there is no such vertex, then $\ell_x = 0$) with edges. The examination happens outwards from this vertex, and count every branch in vertices, which are connected to at least three other vertices with edges. Thus, from the perspective of the original vertex, there are two branches connected to the vertices, aligned outward. Inside branches there may also be other branches etc. Divide the result by the total number of edges in the tree.

The following holds:

$$\ell_- + \ell_x = 1 \quad (15)$$

Examples:

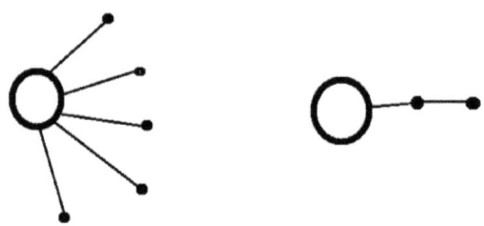

$$\ell_- = \frac{0}{5} = 0 \,, \ell_x = \frac{5}{5} = 1 \qquad \ell_- = \frac{2}{2} = 1 \,, \ell_x = \frac{0}{2} = 0$$

$$\ell_- = \frac{5}{5} = 1 , \ell_x = \frac{0}{5} = 0 \qquad \ell_- = \frac{2}{2} = 1, \ell_x = \frac{0}{2} = 0$$

ℓ_-, ℓ_x not defined

ℓ_-, ℓ_x not defined

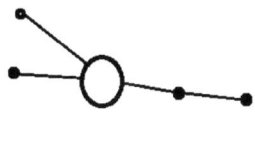

$$\ell_- = \frac{1}{4}, \ell_x = \frac{3}{4}$$

$$\ell_- = \frac{2}{11}, \ell_x = \frac{9}{11}$$

143

The following rule applies to sum forms:

- If $s(i)$ and $s(j)$ are rooted tree sum forms and $s(i) \cong s(j)$, then $\ell_-(s(i)) = \ell_-(s(j))$; $i = j$; $i, j \in \mathbb{N} \setminus \{0\}$.
- If $s(i)$ and $s(j)$ are free tree sum forms and $s(i) \cong s(j)$, then $\ell_-(s(i)) = \ell_-(s(j))$; $i, j \in \mathbb{N}$.

For the case $s(i) \cong s(j), i \neq j$, to be possible for free trees, these sum forms must belong to the modified free tree sequence S'_n , which contains all of the sum forms of the rooted trees sequence S_n , but without the real roots.

For rooted trees,

$s(1) = 1^0$ is the first term in the sequence of all rooted trees.

For free trees,

$s(0) = 1^0 \cdot 0 + 0^{()}$ is the first term in the sequence of all free trees.

Examples about the identity number as a function of sum forms:

$$\ell_-\left(1^o \cdot 0 + 0^{()}\right) = not\ defined$$

$$\ell_-(1^o + 1^- \cdot n) = \begin{cases} not\ defined\,, & n = 0, n = 1 \\ \dfrac{n}{n}\,, & n = 2 \\ \dfrac{0}{n}\,, & n \geq 3 \end{cases}$$

$$\ell_-(1^o + n^x) = \dfrac{0}{n}\,, n \geq 3$$

$$\ell_-(1^o + n^-) = \dfrac{n}{n}\,, n \geq 2$$

$$\ell_-(1^o + n^{mL}) = \begin{cases} \dfrac{0}{n}\,, & n \geq 2m + 1\,, n \geq 3, m = 0 \\ \dfrac{n}{n}\,, & n = 3\,, m = 1 \\ \dfrac{m}{n}\,, & n \geq 2m + 1, n > 3, m \geq 1 \end{cases}$$

$$\ell_-(1^o + n^{mY}) = \dfrac{n-m}{n}\,, n \geq m \geq 3$$

If the identity number ℓ_- for a tree with an index a in the sequence S_n of all trees with a size n is $\ell_-(S_n(a))$, then the average identity number for the sequence S_n is

- $\ell_-(S_n) = \frac{1}{r_n}\sum_{k=1}^{r_n}\ell_-(S_n(k))$, for rooted trees
- $\ell_-(S_n) = \frac{1}{t_n}\sum_{k=1}^{t_n}\ell_-(S_n(k))$, for free trees
- $\ell_-(S_n) = \frac{0}{s_n}$, for series-reduced rooted trees
- $\ell_-(S_n) = \frac{0}{h_n}$, for series-reduced free trees

The interesting problem would be to deduce the average identity numbers for rooted trees and free trees when n approaches infinity, with calculating the limits

- $\lim_{n \to \infty}(\ell_-(S_n)) = \lim_{n \to \infty}\left(\frac{1}{r_n}\sum_{k=1}^{r_n}\ell_-(S_n(k))\right)$
  , for rooted trees
- $\lim_{n \to \infty}(\ell_-(S_n)) = \lim_{n \to \infty}\left(\frac{1}{t_n}\sum_{k=1}^{t_n}\ell_-(S_n(k))\right)$
  , for free trees

The corresponding identity numbers ℓ_x can be calculated with the property

$$\ell_- + \ell_x = 1.$$

2. Create a general method with which you can always generate all the graphically identical forms of a sum form in the order in which they appear in the specific sequence I_{rie-}^{n+a}, without the aid of a computer. $v = n$ is the size of a tree, $-$ means that the sequence is for free trees only (obeys the condition $r_v[-]$), $v' = $ rie and a is the index of the free tree in the sequence S_n. In other words

$$I_{rie-}^{n+a}(1) = S_n(a) = (n)_{rie-}^{ao}.$$

A graphically identical form of the sum form $(n)_{rie-}^{ao}$, with an index k in the sequence I_{rie-}^{n+a} is $I_{rie-}^{n+a}(k)$.

Examples:

$$(3)_{rie-}^{10} = 1^{0-} + 1^- \cdot 2 = I_{rie-}^{3+1}(1)$$

$$\cong 1^{0-} + 2^- = I_{rie-}^{3+1}(2)$$

$$\left|I_{rie-}^{3+1}\right| = 2$$

$$(4)_{rie-}^{1o} = 1^{o-} + 1^- \cdot 3 = I_{rie-}^{4+1}(1)$$

$$\cong 1^{o-} + 3^x = I_{rie-}^{4+1}(2)$$

$$\left| I_{rie-}^{4+1} \right| = 2$$

$$(6)_{rie-}^{3o} = 1^{o-} + 2^- \cdot 2 + 1^- = I_{rie-}^{6+3}(1)$$

$$\cong 1^{o-} + 4^L + 1^- = I_{rie-}^{6+3}(2)$$
$$\cong 1^{o-} + 5^{2L} = I_{rie-}^{6+3}(3)$$
$$\cong 1^{o-} + (1^* + 4^L) = I_{rie-}^{6+3}(4)$$

$$\left| I_{rie-}^{6+3} \right| = 4$$

$$(7)_{rie-}^{8o} = 1^{o-} + 3^- + 2^- + 1^- = I_{rie-}^{7+8}(1)$$

$$\cong 1^{o-} + 4^L + 2^- = I_{rie-}^{7+8}(2)$$
$$\cong 1^{o-} + (1^* + 3^- + 1^-) + 1^- = I_{rie-}^{7+8}(3)$$
$$\cong 1^{o-} + (1^* + 4^L) + 1^- = I_{rie-}^{7+8}(4)$$
$$\cong 1^{o-} + (1^* + 3^- + 2^-) = I_{rie-}^{7+8}(5)$$
$$\cong 1^{o-} + \left(1^* + (1^* + 3^- + 1^-) \right) = I_{rie-}^{7+8}(6)$$
$$\cong 1^{o-} + \left(1^* + (1^* + 4^L) \right) = I_{rie-}^{7+8}(7)$$

$$\left| I_{rie-}^{7+8} \right| = 7 \ .$$

3. Create a method with which you can always deduce the result of $S_+^m(s(i))$ without the aid of a computer, when $s(i)$ is an opened sum form of a rooted tree; $i, m \in \mathbb{N} \setminus \{0\}$.

   This will be the heart of treespeak as a form of art, creative activity, and a competitive game: deducing $S_+^m(s(i))$ for any $i, m \in \mathbb{N} \setminus \{0\}$, when $s(i)$ is an opened sum form of a rooted tree.

4. Create a more general version of the Treespeak game, in which the rules of vertex edge algebra can be applied, to allow many more combinations

5. Create a general symbolic vertex edge algebra version of the Treespeak game

6. Create a player versus player mode version of the Treespeak game